SpringerBriefs in Applied

PoliMI SpringerBriefs

Series Editors

Barbara Pernici, DEIB, Politecnico di Milano, Milano, Italy

Stefano Della Torre, DABC, Politecnico di Milano, Milano, Italy

Bianca M. Colosimo, DMEC, Politecnico di Milano, Milano, Italy

Tiziano Faravelli, DCMC, Politecnico di Milano, Milano, Italy

Roberto Paolucci, DICA, Politecnico di Milano, Milano, Italy

Silvia Piardi, Design, Politecnico di Milano, Milano, Italy

Gabriele Pasqui, DASTU, Politecnico di Milano, Milano, Italy

Springer, in cooperation with Politecnico di Milano, publishes the PoliMI Springer-Briefs, concise summaries of cutting-edge research and practical applications across a wide spectrum of fields. Featuring compact volumes of 50 to 125 (150 as a maximum) pages, the series covers a range of contents from professional to academic in the following research areas carried out at Politecnico:

- Aerospace Engineering
- Bioengineering
- Electrical Engineering
- Energy and Nuclear Science and Technology
- Environmental and Infrastructure Engineering
- Industrial Chemistry and Chemical Engineering
- Information Technology
- Management, Economics and Industrial Engineering
- Materials Engineering
- Mathematical Models and Methods in Engineering
- Mechanical Engineering
- Structural Seismic and Geotechnical Engineering
- Built Environment and Construction Engineering
- Physics
- Design and Technologies
- Urban Planning, Design, and Policy

http://www.polimi.it

Laura A. Pellegrini · Elvira Spatolisano ·
Federica Restelli · Giorgia De Guido ·
Alberto R. de Angelis · Andrea Lainati

Green H$_2$ Transport through LH$_2$, NH$_3$ and LOHC

Opportunities and Challenges

POLITECNICO
MILANO 1863

🐎 Springer

Laura A. Pellegrini
Department of Chemistry, Materials
and Chemical Engineering
Politecnico di Milano
Milan, Italy

Elvira Spatolisano
Department of Chemistry, Materials
and Chemical Engineering
Politecnico di Milano
Milan, Italy

Federica Restelli
Department of Chemistry, Materials
and Chemical Engineering
Politecnico di Milano
Milan, Italy

Giorgia De Guido
Department of Chemistry, Materials
and Chemical Engineering
Politecnico di Milano
Milan, Italy

Alberto R. de Angelis
Research and Technological Innovation
Department
Eni S.p.A.
San Donato Milanese, Italy

Andrea Lainati
Research and Technological Innovation
Department
Eni S.p.A.
San Donato Milanese, Italy

ISSN 2191-530X   ISSN 2191-5318 (electronic)
SpringerBriefs in Applied Sciences and Technology
ISSN 2282-2577   ISSN 2282-2585 (electronic)
PoliMI SpringerBriefs
ISBN 978-3-031-66555-4   ISBN 978-3-031-66556-1 (eBook)
https://doi.org/10.1007/978-3-031-66556-1

© The Editor(s) (if applicable) and The Author(s), under exclusive license to Springer Nature Switzerland AG 2024

This work is subject to copyright. All rights are solely and exclusively licensed by the Publisher, whether the whole or part of the material is concerned, specifically the rights of translation, reprinting, reuse of illustrations, recitation, broadcasting, reproduction on microfilms or in any other physical way, and transmission or information storage and retrieval, electronic adaptation, computer software, or by similar or dissimilar methodology now known or hereafter developed.
The use of general descriptive names, registered names, trademarks, service marks, etc. in this publication does not imply, even in the absence of a specific statement, that such names are exempt from the relevant protective laws and regulations and therefore free for general use.
The publisher, the authors and the editors are safe to assume that the advice and information in this book are believed to be true and accurate at the date of publication. Neither the publisher nor the authors or the editors give a warranty, expressed or implied, with respect to the material contained herein or for any errors or omissions that may have been made. The publisher remains neutral with regard to jurisdictional claims in published maps and institutional affiliations.

This Springer imprint is published by the registered company Springer Nature Switzerland AG
The registered company address is: Gewerbestrasse 11, 6330 Cham, Switzerland

If disposing of this product, please recycle the paper.

Preface

Green hydrogen holds promise as a future energy source, offering a sustainable solution to the world's growing energy needs while simultaneously mitigating climate change. It can be efficiently produced in regions where renewable sources are extensively available, which are typically remote areas far from the large production sites. For this reason, efficient and cost-effective H_2 transport from the production hub to the end users is required. Hydrogen delivery costs significantly affect the economics of its utilization as energy carrier. H_2 has an extremely low volumetric density, such that high compression levels are required for its distribution in gaseous form. In view of addressing the challenges of H_2 transport worldwide, alternative carriers come into play. A H_2 carrier is a substance used to store H_2, facilitating its transport from the renewable energy production hub to the hydrogen utilization terminal. Upon arrival, it can be reconverted to release H_2.

Due to the lack of detailed technical analyses of H_2 transport through different carriers, this work aims at studying different hydrogen transport value chains from the process engineering point of view. Each stage of the value chain is investigated through an in-depth techno-economic assessment, to highlight the critical issues and the need for further investigation (low TRL).

In particular, this work considers:

- the liquefied hydrogen value chain (Chap. 3);
- the ammonia value chain (Chap. 4);
- the toluene/methylcyclohexane value chain (Chap. 5);
- the dibenzyltoluene/perhydrodibenzyltoluene value chain (Chap. 6).

The alternatives are analysed considering:

- different H_2 applications (industrial sector—H_2 valley case and mobility sector—Hydrogen Refuelling Station case);
- different costs of utilities (present and future scenarios);
- different distances from the loading to the unloading terminal, to be covered via ship transport (2500–10000 km).

All these scenarios are discussed in Chap. 7, to understand which is the most cost-effective alternative for each case study.

The main results of the abovementioned hydrogen carriers are summarized below.

Liquefied hydrogen has a volumetric density approximately 800 times greater than that of gaseous hydrogen at atmospheric pressure. Similarly to liquefied natural gas (LNG), liquid hydrogen is obtained through the liquefaction process. However, because of the extremely low H_2 normal boiling point, a higher energy input and a better insulation are required.

According to the techno-economic assessment performed, liquefied hydrogen represents the most disadvantageous alternative as H_2 carrier in the case of industrial applications (H_2 valley), while it is advantageous in the case of mobility sector applications (Hydrogen Refuelling Station—HRS). The cost driver of the value chain is liquefaction. Since operating costs are mostly due to electricity consumption, focusing research efforts on the energy efficiency of the liquefaction process would be beneficial for future optimization.

Ammonia (NH_3) is deemed to be one of the most essential chemicals, normally synthesized from natural gas and mainly used as a building block for the production of fertilizers. However, it can also act as a storage of clean hydrogen, transporting and storing it in refrigerated tanks and splitting it, once it reaches its destination, into its components, N_2 and H_2, through an endothermic cracking process. The advantage of using ammonia as a hydrogen carrier lies in its widespread use, as well as in the possibility of being transported and handled via existing infrastructures dedicated to liquefied petroleum gas (LPG).

The techno-economic analysis carried out demonstrated that the transport and storage of hydrogen using ammonia is advantageous both for industrial applications and in the mobility sector. The cost driver of the chain is the ammonia synthesis stage. Research should therefore be focused on intensifying the Haber-Bosch process to reduce costs, taking into account the smaller scale compared to existing plants.

Toluene Among all the liquid organic hydrogen carriers proposed in literature, toluene shows the highest maturity level, patented by Chiyoda Corporation and available at a demonstration scale. For this reason, it is selected as the benchmark for liquid organic hydrogen carriers (LOHCs). Liquid organic hydrogen carriers are organic molecules which can be reversibly hydrogenated and dehydrogenated to favour H_2 release. Hydrogenation is an exothermic process which consists in chemically reacting the liquid compound with hydrogen, in such a way that the hydrogenated product can be transported at atmospheric pressure. Upon arrival, hydrogen is released via an endothermic dehydrogenation process. The dehydrogenated LOHC can, then, be transported back to the hydrogen source for reuse. The clear advantages of LOHCs lie in their compatibility with existing infrastructure for oil products and their ability to store hydrogen without losses in the long term or during transportation under standard conditions.

From the techno-economic assessment performed, hydrogen transport via LOHCs is advantageous in the case of industrial applications over short distances. The cost driver of the value chain is the dehydrogenation of the carrier for hydrogen release.

Dibenzyltoluene Together with toluene, also dibenzyltoluene has been considered as organic carrier, taking into account its promising physicochemical features. As opposite to toluene, dibenzyltoluene shows a high cost associated to the initial loading required to reach the steady state for the overall cyclic process. The amount of carrier required increases as the distance to be covered via seaborne transport increases. The economic viability of H_2 transport via dibenzyltoluene depends on its future market situation, which in turn depends on alternative processes for its production.

Milan, Italy	Laura A. Pellegrini
Milan, Italy	Elvira Spatolisano
Milan, Italy	Federica Restelli
Milan, Italy	Giorgia De Guido
San Donato Milanese, Italy	Alberto R. de Angelis
San Donato Milanese, Italy	Andrea Lainati

Contents

1 **Green H$_2$: One of the Allies for Decarbonization** 1
   1.1 Introduction 1
   1.2 Green H$_2$: The Key Player for Sustainability 2
   1.3 Green H$_2$ Transport Through H$_2$ Carriers 3
   References 4

2 **Systematic Framework for the Techno-Economic Assessment of Green H$_2$ Value Chains** 7
   2.1 Introduction 7
   2.2 Cost Drivers 8
   2.3 Transport by Ship 12
   2.4 Storage 14
   2.5 Distribution 14
   2.6 Design Bases 16
   References 17

3 **Liquefied H$_2$ as Green H$_2$ Carrier** 19
   3.1 Introduction 19
   3.2 Hydrogen Liquefaction 20
   3.3 Liquefied H$_2$ Regasification 23
      3.3.1 Centralized Regasification 23
      3.3.2 Decentralized Regasification 24
   3.4 Techno-Economic Assessment 25
   References 30

4 **Ammonia as Green H$_2$ Carrier** 33
   4.1 Introduction 33
   4.2 Ammonia Synthesis 34
   4.3 Ammonia Cracking 37

	4.3.1 Centralized Cracking	38
	4.3.2 Decentralized Cracking	39
4.4	Techno-Economic Assessment	41
References	47

5 Toluene/methylcyclohexane as Green H$_2$ Carrier
5.1	Introduction	49
5.2	Toluene Hydrogenation	50
5.3	Methylcyclohexane Dehydrogenation	51
5.4	Techno-Economic Assessment	52
References		57

6 Dibenzyltoluene/Perhydro-Dibenzyltoluene as Green H$_2$ Carrier
6.1	Introduction	59
6.2	Dibenzyltoluene Hydrogenation	60
6.3	Perhydro-Dibenzyltoluene Dehydrogenation	62
6.4	Techno-Economic Assessment	62
References		67

7 Comparison and Future Perspectives
7.1	Introduction	69
7.2	H$_2$ Valley Case	70
7.3	HRS Case	70
7.4	Comparison with the Literature	75
7.5	Conclusions and Future Perspectives	83
References		84

Abbreviations

ASU	Air Separation Unit
BFD	Block Flow Diagram
BFW	Boiler Feed Water
CAPEX	CAPital EXpenses
CEPCI	Chemical Engineering Plant Cost Index
CW	Cooling Water
DBT-H0	DiBenzylToluene
DBT-H18	Perhydrodibenzyltoluene
DMC	Direct Manufacturing Costs
FCI	Fixed Capital Investment
FMC	Fixed Manufacturing Costs
GE	General Expenses
IFO	Intermediate Fuel Oil
LCOH	Levelised Cost Of Hydrogen
LH_2	Liquefied Hydrogen
LNG	Liquefied Natural Gas
LOHC	Liquid Organic Hydrogen Carrier
LPG	Liquefied Petroleum Gas
MCH	MethylCycloHexane
MR	Mixed Refrigerant
OPEX	OPerating EXpenses
PFD	Process Flow Diagram
TOL	TOLuene
TRL	Technology Readiness Level
WACC	Weighted Average Cost of Capital

Chapter 1
Green H$_2$: One of the Allies for Decarbonization

Abstract Green hydrogen is emerging as the pivotal player in the journey toward the net-zero-emissions global economy. Nevertheless, its transportation over long distances at a large scale is challenging considering both costs and technology, primarily due to H$_2$ low volumetric energy density. To delve into these challenges, different hydrogen carriers can be exploited, such as liquid hydrogen, ammonia, or liquid organic hydrogen carriers. Hydrogen carriers serve as a medium to store and transport hydrogen in a safe and efficient manner, enabling its delivery from the production sites to the end users. Upon arrival at the end users, these carriers can undergo reconversion for gaseous H$_2$ release. Various hydrogen carriers exist, each with distinct features and suitability for different transport modes and applications. These carriers undergo continuous innovation and development to enhance their efficiency, safety, and economic viability. To help understanding their applicability, opportunities and drawbacks of each of them are pointed out in this section.

Keywords H$_2$ carriers · H$_2$ transport · Decarbonization · Net-zero economy · Energy density

1.1 Introduction

Numerous nations across the globe are actively working to reduce their dependency on non-renewable and polluting energy sources like coal, oil, and gas. This endeavour is bolstered by governmental efforts aimed at increasing the utilization of renewable energy, particularly solar and wind power. The Renewable Energy Policy Network for the 21st Century (REN) witnessed the commitment of 61 countries to decrease their contribution to climate change by enhancing the proportion of renewable energy in electricity generation [1]. The November 2015 Paris Agreement set the ambitious goal of eliminating net greenhouse gas emissions by the latter half of this century. However, the availability of renewable energy resources varies significantly across different regions, resulting in disparities between energy consumption levels and resource accessibility. While certain regions possess abundant potential for green energy, others encounter limitations. This evidence underscores the necessity for a diverse array of renewable energy sources and technologies to meet global energy demands. Consequently, as the world shifts towards a greater reliance on wind and solar power, there is an anticipated need for expanded energy transportation over long distances. Furthermore, the adoption of renewable energy infrastructure entails substantial investments in research, development, and deployment, highlighting the importance of collaboration and support from governments, industry, and society at large.

Renewable energy generation, exemplified by wind and photovoltaic sources, inherently fluctuates due to meteorological conditions. Consequently, any energy system heavily dependent on such intermittent sources requires storage capabilities to align stochastic production with specific energy demands. Therefore, energy storage systems are vital for storing surplus energy generated during peak production periods and discharging it during low-production phases. The importance of large-scale energy storage thus becomes paramount [2].

1.2 Green H$_2$: The Key Player for Sustainability

As countries worldwide commit to reducing their reliance on fossil fuels and transitioning to a low-carbon economy, green hydrogen emerges as a key player in achieving these goals, offering a clean, sustainable, and scalable energy solution for the future. Its versatility as an energy source and feedstock is evident across various industries, particularly those where direct electrification is unfeasible. Unlike fuels such as diesel and gasoline, the combustion of hydrogen emits no pollutants like NO_x and SO_x, rendering it an exceptionally clean fuel option. Hydrogen combustion results in the generation of water and releases a net heat energy of approximately 120 MJ/kg. Notably, hydrogen boasts an impressive gravimetric energy storage density, with each kilogram containing roughly 33 kWh of energy. A comparison of different fuels is presented in Fig. 1.1 for reference.

Once produced exploiting renewable sources, H$_2$ has to be delivered to the end user. The efficient transportation of green hydrogen from production facilities to end users at minimal cost is essential for its widespread adoption. Hence, exploring diverse transportation solutions is imperative to identify the most practical options.

Transportation costs mainly depend on two factors: the volume of hydrogen being transported and the distance it needs to travel. There are three main infrastructures for transporting hydrogen: by truck, pipeline, or ship. Trucks are suitable for small volumes and short distances, but as volume increases, the need for higher density and the use of liquid hydrogen-carrying trucks becomes necessary. As volume or distance increases further, pipelines become the most cost-effective option, starting with smaller distribution pipelines and progressing to larger transmission pipelines. For long distances or routes across bodies of water, ships are the preferred choice. Therefore, in the context of cross-border trade, trucks are typically not preferred due to their suitability for smaller volume shipments [4].

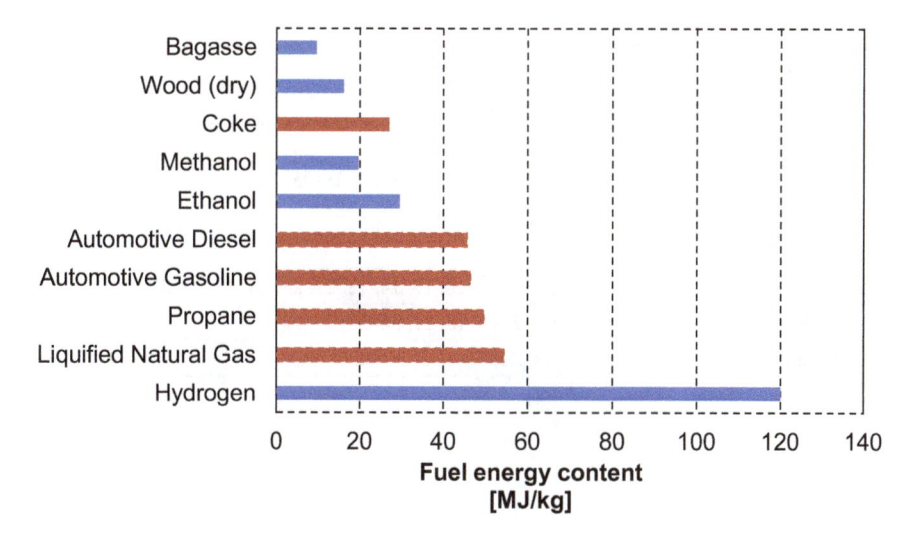

Fig. 1.1 Energy content of different fuels. Red ones are fossil-based [3]

1.3 Green H$_2$ Transport Through H$_2$ Carriers

Together with the most suitable infrastructure, to allow for a cost-effective transport, H$_2$ can be either liquefied or transformed into various hydrogen-based fuels and feedstocks like synthetic methane, liquid fuels and ammonia, leveraging existing infrastructure for their distribution, thus potentially reducing costs for end users. Some of these synthetic hydrocarbons can directly substitute their fossil counterparts [5]. Ammonia, already utilized as a feedstock in the chemical industry, holds promise as a carrier for long-distance hydrogen transport or as a fuel in the shipping sector.

Several hydrogen carriers have been proposed in literature. Many of them show a very low technology readiness level (TRL) and require a lot of research effort before their application on a large scale.

An overview of the most recent techno-economic assessments of hydrogen transport through different carriers is reported in Table 1.1.

Different studies yield varied results depending on their underlying hypotheses, making it challenging to draw overarching conclusions. Most literature aims to showcase the economic viability of carriers for hydrogen transport rather than providing a technical assessment of each stage of the H$_2$ value chain.

From the process engineering perspective, it is of interest to focus on H$_2$ conversion to the carrier and the carrier reconversion to H$_2$, i.e., the cost-drivers of the whole H$_2$ transport value chain, to highlight weaknesses and try to address challenges.

Accordingly, in the following chapters a systematic methodology for analyzing various hydrogen value chains is presented. The methodology is then applied to different H$_2$ carriers, to offer a peer comparison and point out opportunities for future process intensification.

Table 1.1 Most recent techno-economical evaluations on the H_2 value chain, classified according to the H_2 carrier investigated

H_2 carrier investigated	References
Liquefied H_2	[2, 5–26]
Ammonia	[2, 5–10, 12–17, 19, 20, 23–25, 27–32]
Methanol	[15, 16, 18, 28, 33]
Formic acid	[30, 34]
Toluene	[2, 5, 7, 8, 10, 12, 14, 18, 21, 23, 28, 30, 35–37]
Dibenzyltoluene	[9, 12, 16, 18, 21, 36–38]
N-ethylcarbazole	[18, 22]
Liquefied Natural Gas	[23]

References

1. Primer GSR. Renewable energy policy network for the 21st century. 2009.
2. Hydrogen transportation. 2021.
3. The Royal Society. Ammonia: zero-carbon fertiliser, fuel and energy store. 2020.
4. IRENA. Global hydrogen trade to meet the 1.5 °C climate goal: Part II—Technology review of hydrogen carriers;2022. Abu Dhabi: International Renewable Energy Agency
5. The Future of Hydrogen. 2019.
6. Okunlola A, Giwa T, Di Lullo G, Davis M, Gemechu E, Kumar A. Techno-economic assessment of low-carbon hydrogen export from Western Canada to Eastern Canada, the USA, the Asia-Pacific, and Europe. Int J Hydrogen Energy. 2022;47:6453–77.
7. Hong X, Thaore VB, Karimi IA, Farooq S, Wang X, Usadi AK, et al. Techno-enviro-economic analyses of hydrogen supply chains with an ASEAN case study. Int J Hydrogen Energy. 2021;46:32914–28.
8. Wijayanta AT, Oda T, Purnomo CW, Kashiwagi T, Aziz M. Liquid hydrogen, methylcyclohexane, and ammonia as potential hydrogen storage: Comparison review. Int J Hydrogen Energy. 2019;44:15026–44.
9. Di Lullo G, Giwa T, Okunlola A, Davis M, Mehedi T, Oni AO, et al. Large-scale long-distance land-based hydrogen transportation systems: a comparative techno-economic and greenhouse gas emission assessment. Int J Hydrogen Energy. 2022;47:35293–319.
10. Johnston C, Ali Khan MH, Amal R, Daiyan R, MacGill I. Shipping the sunshine: an open-source model for costing renewable hydrogen transport from Australia. Int J Hydrogen Energy. 2022;47:20362–77.
11. Correa G, Volpe F, Marocco P, Muñoz P, Falagüerra T, Santarelli M. Evaluation of levelized cost of hydrogen produced by wind electrolysis: Argentine and Italian production scenarios. J Energy Storage. 2022;52: 105014.
12. Lee J-S, Cherif A, Yoon H-J, Seo S-K, Bae J-E, Shin H-J, et al. Large-scale overseas transportation of hydrogen: comparative techno-economic and environmental investigation. Renew Sustain Energy Rev. 2022;165: 112556.
13. Sens L, Neuling U, Wilbrand K, Martin K. Conditioned hydrogen for a green hydrogen supply for heavy duty-vehicles in 2030 and 2050—a techno-economic well-to-tank assessment of various supply chains. Int J Hydrogen Energy. 2022.
14. Song S, Lin H, Sherman P, Yang X, Nielsen CP, Chen X, et al. Production of hydrogen from offshore wind in China and cost-competitive supply to Japan. Nat Commun. 2021;12:6953.
15. Gallardo FI, Ferrario AM, Lamagna M, Bocci E, Garcia DA, Baeza-Jeria TE. A Techno-Economic Analysis of solar hydrogen production by electrolysis in the north of Chile and the case of exportation from Atacama Desert to Japan. Int J Hydrogen Energy. 2021;46:13709–28.

16. Hank C, Sternberg A, Köppel N, Holst M, Smolinka T, Schaadt A, et al. Energy efficiency and economic assessment of imported energy carriers based on renewable electricity. Sustain Energy Fuels. 2020;4:2256–73.
17. Ishimoto Y, Voldsund M, Nekså P, Roussanaly S, Berstad D, Gardarsdottir SO. Large-scale production and transport of hydrogen from Norway to Europe and Japan: Value chain analysis and comparison of liquid hydrogen and ammonia as energy carriers. Int J Hydrogen Energy. 2020;45:32865–83.
18. Niermann M, Timmerberg S, Drünert S, Kaltschmitt M. Liquid Organic Hydrogen Carriers and alternatives for international transport of renewable hydrogen. Renew Sustain Energy Rev. 2021;135:110171.
19. Song Q, Tinoco RR, Yang H, Yang Q, Jiang H, Chen Y, et al. A comparative study on energy efficiency of the maritime supply chains for liquefied hydrogen, ammonia, methanol and natural gas. Carbon Capture Sci Technol. 2022;4:100056.
20. Kim J, Huh C, Seo Y. End-to-end value chain analysis of isolated renewable energy using hydrogen and ammonia energy carrier. Energy Conv Manage. 2022;254:115247.
21. Raab M, Maier S, Dietrich R-U. Comparative techno-economic assessment of a large-scale hydrogen transport via liquid transport media. Int J Hydrogen Energy. 2021;46:11956–68.
22. Rong Y, Chen S, Li C, Chen X, Xie L, Chen J, et al. Techno-economic analysis of hydrogen storage and transportation from hydrogen plant to terminal refueling station. Int J Hydrogen Energy. 2024;52:547–58.
23. Chodorowska N, Farhadi, M. H_2 value chain comparing different transport vectors. In: IGPA Europe virtual conference. 25 May 2021.
24. Seo Y, Park H, Lee S, Kim J, Han S. Design concepts of hydrogen supply chain to bring consumers offshore green hydrogen. Int J Hydrogen Energy. 2023.
25. Ibrahim Y, Al-Mohannadi DM. Optimization of low-carbon hydrogen supply chain networks in industrial clusters. Int J Hydrogen Energy. 2023;48:13325–42.
26. Restelli F, Spatolisano E, Pellegrini LA, Cattaneo S, de Angelis AR, Lainati A, et al. Liquefied hydrogen value chain: a detailed techno-economic evaluation for its application in the industrial and mobility sectors. Int J Hydrogen Energy. 2024;52:454–66.
27. The Future of Hydrogen. Seizing today's opportunities. Report prepared by the IEA for the G20, Japan;2019. Agency IE.
28. Papadias DD, Peng J-K, Ahluwalia RK. Hydrogen carriers: Production, transmission, decomposition, and storage. Int J Hydrogen Energy. 2021;46:24169–89.
29. Makhloufi C, Kezibri N. Large-scale decomposition of green ammonia for pure hydrogen production. Int J Hydrogen Energy. 2021;46:34777–87.
30. Crandall BS, Brix T, Weber RS, Jiao F. Techno-economic assessment of green H_2 carrier supply chains. Energy Fuels. 2023;37:1441–50.
31. Cui J, Aziz M. Chemical engineering transactions economic assessment of green hydrogen infrastructure: a case study in China. 2023;103:313–8.
32. Restelli F, Spatolisano E, Pellegrini LA, de Angelis AR, Cattaneo S, Roccaro E. Detailed techno-economic assessment of ammonia as green H_2 carrier. Int J Hydrogen Energy. 2024;52:532–47.
33. Hydrogen Import Coalition. Shipping sun and wind to Belgium is key in climate neutral economy. Technology. 2021;14:19.
34. Kim C, Lee Y, Kim K, Lee U. Implementation of formic acid as a Liquid Organic Hydrogen Carrier (LOHC): techno-economic analysis and life cycle assessment of formic acid produced via CO_2 utilization. Catalysts;2022.
35. Noh H, Kang K, Seo Y. Environmental and energy efficiency assessments of offshore hydrogen supply chains utilizing compressed gaseous hydrogen, liquefied hydrogen, liquid organic hydrogen carriers and ammonia. Int J Hydrogen Energy. 2023;48:7515–32.
36. Godinho J, Hoefnagels R, Braz CG, Sousa AM, Granjo JFO. An economic and greenhouse gas footprint assessment of international maritime transportation of hydrogen using liquid organic hydrogen carriers. Energy. 2023.

37. Spatolisano E, Restelli F, Matichecchia A, Pellegrini LA, de Angelis AR, Cattaneo S, et al. Assessing opportunities and weaknesses of green hydrogen transport via LOHC through a detailed techno-economic analysis. Int J Hydrogen Energy. 2024;52:703–17.
38. Hurskainen M, Ihonen J. Techno-economic feasibility of road transport of hydrogen using liquid organic hydrogen carriers. Int J Hydrogen Energy. 2020;45:32098–112.

Chapter 2
Systematic Framework for the Techno-Economic Assessment of Green H$_2$ Value Chains

Abstract Despite its wide application as a chemical on industrial scale, hydrogen utilization as an energy vector still suffers from unfavourable economics, mainly due to its high cost of production, storage and transportation. To overcome the last two of these issues, different hydrogen carriers have been proposed. Hydrogen storage and transportation through these carriers involve: 1. the conversion of green H$_2$, produced at the loading terminal where renewable sources are easily accessible, to the considered carrier, 2. the storage and transportation of the liquid carrier and 3. its subsequent reconversion to H$_2$ at the unloading terminal. Although there is a number of studies in literature on the economic feasibility of hydrogen transport through different H$_2$ vectors, very few of them delve into the technical evaluation of the hydrogen value chain. From the process design point of view, the conversion and reconversion stages are of paramount importance, considering that they are the cost drivers of the whole system. This work aims to address this gap by presenting a systematic methodology to technically analyse different hydrogen vectors. The systematic framework pointed out can be applied to each H$_2$ carrier considered, in view of assessing opportunities and challenges for each of them.

Keywords *CAPEX* and *OPEX* · Levelised cost of H$_2$ transport · Harbour-to-Harbour H$_2$ transport · Techno-economic assessment · Systematic methodology · H$_2$ valley · H$_2$ refuelling stations

2.1 Introduction

The different analysed hydrogen carriers, i.e., liquefied hydrogen (LH$_2$), ammonia (NH$_3$) and liquid organic hydrogen carriers (LOHCs), are compared with each other through techno-economic evaluations. As can be easily understood, the cost drivers of each chain are the hydrogen conversion stages in the carrier considered and vice versa.

For the economic evaluation of the cost drivers of each value chain (liquefaction and regasification of hydrogen, synthesis and cracking of ammonia, hydrogenation and dehydrogenation of organic carriers), the fixed and operating costs are estimated using the Turton methodology [1], starting from the process simulations detailed in the next chapters. On the other hand, the economic evaluation for the storage, shipping and distribution sections is based on data found in the literature, as explained in the following paragraphs.

Once the capital expenditures (*CAPEX*) and operating expenditures (*OPEX*) for each section of the chain are known, the Levelised Cost of Hydrogen (*LCOH*) [€/kg$_{H2}$] is calculated using relation (2.1). The *LCOH* is a common metrics for the benchmark of the cost-competitiveness of hydrogen production chains. It is a variable which indicates the cost per kg of green hydrogen, taking into account the estimated investment and operating costs for the assets involved in the considered scenario. *LCOH* is given by the following equation:

$$LCOH = \frac{\sum_{t=0}^{n-1}(CAPEX_t + OPEX_t)(1+r)^{-t}}{\sum_{t=0}^{n-1} P_{H2}(1+r)^{-t}} \qquad (2.1)$$

where *LCOH* is the levelized cost of hydrogen (€/kg$_{H2}$ delivered)

P_{H2} is the annual amount of hydrogen delivered,

r is the discount rate, assumed equal to 5%,

t is the year, where $t = 0$ is the base year (2022) and $n - 1$ is the end year, with n being the plant lifetime, assumed to be 25 years,

$CAPEX_t$ is the capital expenditures at time t,

$OPEX_t$ is the operational expenditures at time t.

2.2 Cost Drivers

The Turton methodology for economic analysis allows a rough estimate of plant costs and has to be intended as a preliminary feasibility study for the application of the technology on an industrial scale.

The methodology involves, for each piece of equipment, the calculation of the purchased base cost ($C_{p,i}^0$), using the equation:

$$\log_{10}\left(C_{p,i}^0(2001)\right) = K_{1,i} + K_{2,i}\log_{10}(A_i) + K_{3,i}\left[\log_{10}(A_i)\right]^2 \qquad (2.2)$$

In Eq. (2.2), A_i is a characteristic size of equipment and depends on the type of equipment. The constants $K_{1,i}$, $K_{2,i}$, $K_{3,i}$ to be used in relation (2.2) are available in Turton et al. [1] and allow to evaluate the cost of the equipment referring to the year 2001. These costs must be referred to the base year for the economic evaluation, taking into account economic inflation by means of the following expression:

$$C_2 = C_1\left(\frac{I_2}{I_1}\right) \qquad (2.3)$$

2.2 Cost Drivers

being: C purchased cost;
I cost index;
subscript 1 year to which the calculated cost C refers;
subscript 2 year to which the current C cost refers.

The cost index used in this analysis is the Chemical Engineering Plant Cost Index (CEPCI), which is CEPCI (2001) = 397 and CEPCI (2022) = 816.5 [2].

The purchased base cost, however, does not consider the effect on the equipment cost of the operating pressure and the material of construction, which must be introduced by calculating the bare module cost ($C_{BM,i}$) as:

$$C_{BM,i} = C^0_{p,i} F_{BM,i} = C^0_{p,i} \left(B_{1,i} + B_{2,i} F_{M,i} F_{P,i} \right) \tag{2.4}$$

where $F_{BM,i}$ (bare module factor) is a function of the parameters $F_{P,i}$ (pressure factor), which considers the operating pressure of the equipment, and $F_{M,i}$ (material factor), which depends on the material of construction. $B_{1,i}$ and $B_{2,i}$, are constants specific for each equipment i and can be found in Turton et al. [1].

The bare module cost for atmospheric operating pressure and carbon steel construction material, and the bare module factor for the equipment under these conditions are calculated by setting $F_{M,i} = F_{P,i} = 1$.

The total module cost (C_{TM}) is calculated, starting from the bare module cost for the single equipment, using Eq. (2.5) and is used for determining the CAPEX as in Eq. (2.6). This expression considers the cost of a green field plant (grassroot cost, C_{GR}) by appropriately increasing the total module cost.

$$C_{TM} = 1.18 \sum_{i=1}^{n} C_{BM,i} \tag{2.5}$$

$$CAPEX = C_{GR} = C_{TM} + 0.5 \sum_{i=1}^{n} C^0_{BM,i} \tag{2.6}$$

Similarly to what is done for the fixed costs, the operating costs for the cost drivers of each of the value chains are also calculated according to the methodology presented in Turton et al. [1].

This methodology involves the calculation of the OPEX, considering the direct costs of manufacturing (DCM), the fixed costs of manufacturing (FCM) and the general expenses (GE).

The OPEX are determined as detailed in Table 2.1 from the following quantities:

- CAPEX;
- labour cost (C_{OL});
- cost of utilities (C_{UT});
- cost of waste treatment (C_{WT});
- cost of raw materials (C_{RM}).

Table 2.1 Estimation of operating costs according to the Turton methodology [1]

Cost item	Description	Computed as
Direct Costs of Manufacturing (DCM)		
Raw materials (C_{RM})	Cost of the raw materials	See Table 2.3
Waste treatment (C_{WT})	Cost of waste treatment to comply with environmental limitations	Neglected
Utilities (C_{UT})	Cost of necessary utilities (electricity, steam, cooling water, fuel)	See Table 2.2
Operating labour (C_{OL})	Cost of operating personnel in the plant	See Eqs. (2.7)–(2.11)
Direct supervisory and clerical labour	Cost of support, administrative and engineering personnel	$0.18 \cdot C_{OL}$
Maintenance and repairs	Cost of maintenance and repairs	$0.06 \cdot CAPEX$
Operating supplies	Cost of material not considered in raw materials but necessary for daily operations (paper, filters, respirators, PPE)	$0.009 \cdot CAPEX$
Laboratory charges	Cost of laboratory tests necessary for quality control and diagnosis of malfunctions	$0.15 \cdot C_{OL}$
Patents and royalties	Cost for the use of proprietary technologies	$0.03 \cdot OPEX$
Total (DCM)		$DCM = C_{RM} + C_{WT} + C_{UT} + 1.33 \cdot C_{OL} + 0.069 \cdot CAPEX + 0.03 \cdot OPEX$
Fixed Costs of Manufacturing (FCM)		
Depreciation		Neglected
Local taxes and insurance	Cost associated with insurance and taxes, linked to the location of the plant	$0.032 \cdot CAPEX$
Plant overhead costs	Cost for auxiliary structures (accounting and salary services, health services, canteen)	$0.708 \cdot C_{OL} + 0.036 \cdot CAPEX$
Total (FCM)		$FMC = 0.708 \cdot C_{OL} + 0.068 \cdot CAPEX + depreciation$
General Expenses (GE)		
Administration costs	Administration cost (salaries, facilities and other related activities)	$0.177 \cdot C_{OL} + 0.009 \cdot CAPEX$
Distribution and selling costs	Costs for distribution and selling of chemicals	$0.11 \cdot OPEX$
Research and development	Cost of research and similar activities	$0.05 \cdot OPEX$
Contingency		$0.05 \cdot OPEX$
Total (GE)		$GE = 0.177 \cdot C_{OL} + 0.009 \cdot CAPEX + 0.16 \cdot OPEX$

2.2 Cost Drivers

Table 2.2 Unit cost of utilities for the "present" and "future" scenarios

		"Present"	"Future"
CO_2 emissions[1]	€/t	90	105
Electric energy	€/MWh	500	220
Cooling water (30–40 °C)	€/t	0.015	0.015
Boiler feed water	€/t	1.15	1.20
LP steam	€/t	100	40
MP steam (200 °C, 15 bara)	€/t	130	50
IFO 380 1%S	€/t	580	450
Diesel	€/l	1.8156	1.8156

Table 2.3 Unit cost of raw materials for the "present" and "future" scenarios

		"Present"	"Future"
Toluene	€/t	1300	850
Dibenzyltoluene	€/t	5000	3000
Nitrogen	€/Nm³	0.20	0.15

The C_{OL} is evaluated considering the number of operators needed at the same time in the plant ($N_{OP,plant}$), based on the required shifts. The number of operators per shift, N_{OL}, is calculated as follows:

$$N_{OL} = \left(6.29 + 31.7 \cdot P^2 + 0.23 \cdot N_{np}\right)^{0.5} \tag{2.7}$$

with P being the number of units involving solid particulate management (e.g. transportation and distribution, particle size control), and N_{np}, the number of equipment handling fluids without pumps and valves, evaluated as:

$$N_{np} = \sum_{\substack{compressors \\ columns \\ reactors \\ heat\ exchangers}} units \tag{2.8}$$

$$N_{OP,plant} = N_{OP,necessary} \cdot N_{OL} \tag{2.9}$$

$$N_{OP,necessary} = \frac{N_{shifts/year}\left[\frac{shifts}{y}\right]}{N_{shifts/operator\ year}\left[\frac{shifts}{op \cdot y}\right]} = \frac{8000\left[\frac{h}{y}\right]}{8\left[\frac{h}{shift}\right] \cdot 45\left[\frac{week}{y}\right] \cdot 5\left[\frac{shifts}{op \cdot week}\right]} \tag{2.10}$$

---

[1] The cost associated with CO_2 emissions is equal to the price of emission quotas (Emission Trading System) – EUA, European Union Allowances.

Considering an annual labour cost per operator equal to approximately 40000 €/(y·op), the C_{OL} is determined as:

$$C_{OL} = N_{OP,plant} \cdot 40000 \left[\frac{€}{y \cdot op}\right] \qquad (2.11)$$

As regards the C_{UT} and the C_{RM}, reference is made to the cost items reported, respectively, in Tables 2.2 and 2.3. Due to the significant inflation for the year 2022, two different scenarios are analysed: a first scenario, "present", which takes into account the cost of utilities in the year 2022 and a second one, "future", which takes into account the decrease in prices expected over 4 years. It can be noted that the cost of electricity is particularly high in the present scenario.

Furthermore, for each of the described value chains, the following assumptions are introduced:

- initial and make-up costs of the catalyst neglected;
- spare units neglected;
- port infrastructure neglected;
- only direct CO_2 emissions considered.

2.3 Transport by Ship

For ship transport of the analysed carrier, the fixed costs related to the ship purchase and the operating costs related to the labour, maintenance and insurance, fuel and CO_2 emissions are considered.

As regards the fixed costs, the purchase of one single ship is assumed, whose size is calculated based on the number of days in which the product must be stored ($N°_{days\ of\ prod.\ to\ store}$) defined as follows:

$$N°_{days\ of\ prod.\ to\ store} = t_{go\text{-and-back}} + t_{loading\ and\ unloading} + t_{safety\ margin} \qquad (2.12)$$

where: $t_{return\ voyage}$ is the number of days for the round trip;
$t_{loading\ and\ unloading}$ is the number of days needed for the loading and unloading operations, assumed equal to 1 day;
$t_{safety\ margin}$ is the extra-time to cover any delays, assumed to be 2 days.

Considering, for the ship, a speed of 16 knots (about 30 km/h), the time to make a round trip is calculated using relation (2.13), as a function of the harbour-to-harbour distance D varying in the range 2500–10000 km.

$$t_{go\text{-and-back}}[d] = \frac{2 \cdot D\,[km]}{30\,[km/h] \cdot 24\,[h/d]} \qquad (2.13)$$

2.3 Transport by Ship

Table 2.4 Number of days of production to store $N^\circ_{\text{days of prod. to store}}$ [d] as the harbour-to-harbour distance D [km] varies

D (km)	$t_{\text{go-and-back}}$ [d]	$N^\circ_{\text{days of prod. to store}}$ [d]
2500	7	10
5000	14	17
10000	28	31

Table 2.4 shows the number of days of production to be stored as the harbour-to-harbour distance D varies.

The net capacity of the ship is calculated supposing that a maximum of 98% of its volume can be used. In the case of transport of refrigerated liquids (i.e., liquefied hydrogen and ammonia), it is assumed that approximately 4% of the volume must remain in the tanks to keep them at temperature. Therefore, the capacity of the vessel (V_{vessel}) is calculated via Eq. (2.14) in the case of LOHCs and Eq. (2.15) in the case of liquefied hydrogen/ammonia.

$$V_{\text{vessel}}\left[m^3\right] = \frac{\text{Prod.}[\text{kg}/\text{d}] \cdot N^\circ_{\text{days of prod. to store}}[\text{d}]}{\rho_{\text{species}}[\text{kg}/m^3] \cdot (0.98)} \tag{2.14}$$

$$V_{\text{vessel}}\left[m^3\right] = \frac{\text{Prod.}[\text{kg}/\text{d}] \cdot N^\circ_{\text{days of prod. to store}}[\text{d}]}{\rho_{\text{species}}[\text{kg}/m^3] \cdot (0.98 - 0.04)} \tag{2.15}$$

In Eqs. (2.14)–(2.15), ρ_{species} is the volumetric density of the carrier and Prod. is the daily production of the carrier.

The cost of labour is calculated, for a specific ship type, retrieving the crew size from literature and considering that two complete crews are required in a year, whose single operator has a salary of 52000 €/(y·op).

It is assumed that the propulsion of the ship occurs with a traditional engine having Intermediate Fuel Oil (IFO 380) as fuel. When available, the fuel consumption ($Cons_{\text{fuel}}$ in 2.16) is taken from the technical data sheet of the ship itself. If this information is not provided, the consumption of IFO 380 is calculated given the ship's engine power, assuming a specific consumption of 0.1587 t/MWh [3]. The operating costs related to fuel consumption are, therefore, calculated as:

$$OPEX_{\text{fuel}}\left[\frac{M€}{y}\right] = Cons_{\text{fuel}}\left[\frac{t_{\text{IFO 380}}}{h}\right] \cdot C_{\text{fuel}}\left[\frac{M€}{t_{\text{IFO 380}}}\right] \cdot \frac{t_{\text{go-and-back}}[\text{d}]}{N^\circ_{\text{days of prod. to store}}[\text{d}]} \cdot 8000\left[\frac{h}{y}\right] \tag{2.16}$$

The cost associated with CO_2 emissions is calculated assuming specific CO_2 emissions for IFO 380 (e_{fuel}) equal to 11.24 kg$_{CO_2}$/gallon$_{\text{IFO-380}}$ and volumetric density ($\rho_{\text{IFO 380}}$) equal to 990 kg/m^3:

$$OPEX_{CO_2 \text{ emissions}}\left[\frac{M€}{y}\right] = E_{\text{fuel}}\left[\frac{t_{CO_2}}{h}\right] \cdot C_{CO_2 \text{ emissions}}\left[\frac{M€}{t_{CO_2}}\right] \cdot \frac{t_{\text{go-and-back}}[\text{d}]}{N^\circ_{\text{days of prod. to store}}[\text{d}]} \cdot 8000\left[\frac{h}{y}\right] \tag{2.17}$$

where E_{fuel} is calculated as:

$$E_{fuel}\left[\frac{tCO_2}{h}\right] = \frac{Cons_{fuel}[t_{IFO\,380}/h] \cdot e_{fuel}[kg_{CO_2}/gallon_{IFO\,380}] \cdot 264.2[gallon/m^3]}{\rho_{IFO\,380}[kg/m^3]} \quad (2.18)$$

Maintenance and insurance costs are calculated as 10% of the *CAPEX*.

2.4 Storage

For the storage of the considered carrier, the fixed costs related to the tanks and the *OPEX* related to maintenance and insurance, accounted as 10% of the *CAPEX*, are considered. Other operating costs are neglected, which rigorously should take into account labour costs for loading and unloading operation together with the cost of the electricity needed by the pumps.

The fixed costs associated with the tank are found in literature. For each terminal, the purchase of one single tank is assumed, whose size is equal to the volume transported by the vessel V_{vessel} increased by 10% as a safety margin to take into account any delay.

$$V_{tank}[m^3] = (1 + 0.1) \cdot V_{vessel}[m^3] \quad (2.19)$$

In the case of LOHCs it is necessary to double the tanks due to the need of the simultaneous storage of both the hydrogenated and the dehydrogenated compounds. Fixed costs related to loading pumps and port infrastructure (pier, mooring, H_2 flare, pipelines, etc.) are neglected in this analysis.

2.5 Distribution

As regards the distribution of the considered carrier, the costs associated with the trucks purchase and the operating costs related to labour, fuel and CO_2 emissions are considered. Costs of maintenance and insurance are accounted as 10% of the *CAPEX*.

The fixed costs associated with the single trailer truck C_{truck} are found in the literature. As in Eq. (2.20), they are the sum of the cost of the tractor (C_{engine}) and of the cost of the trailer ($C_{trailer}$):

$$C_{truck} = C_{engine} + C_{trailer} \quad (2.20)$$

2.5 Distribution

where the cost of the tractor can be assumed to be the same for all the transported carriers and equal to 0.29 M€ ([4], cost adjusted for inflation in 2022).

The number of trucks to purchase n_{trucks} is calculated considering that each truck makes 2 trips per day:

$$n_{trucks} = \frac{V_{unloaded\ from\ the\ vessel}[m^3]}{V_{truck}[m^3] \cdot N^\circ_{days\ of\ prod.\ to\ store}[d] \cdot 2[d^{-1}]} \quad (2.21)$$

where V_{truck} is the net capacity of a single truck trailer, and $V_{unloaded\ from\ the\ vessel}$ is calculated accounting for the losses due to the boil-off phenomenon, occurring at a rate equal to r_{BOG}:

$$V_{unloaded\ from\ the\ vessel} = \frac{Prod.[kg/d] \cdot N^\circ_{days\ of\ prod.\ to\ store}[d]}{\rho_{species}[kg/m^3]} \cdot (1 - r_{BOG}[d^{-1}])^{t_{go}[d]}$$

$$(2.22)$$

As regards the operating costs, the cost of labour is determined considering a number of drivers per shift equal to the number of trucks. Each operator has a salary of 40000 €/(y·op).

It is assumed that the trucks have a diesel engine and consume $Cons_{fuel} = 35$ L/100 km [5]. The operating costs associated with fuel consumption are therefore calculated as:

$$OPEX_{fuel}\left[\frac{M€}{y}\right] = Cons_{fuel}\left[\frac{L}{km}\right] \cdot d[km] \cdot 2 \cdot \frac{C_{fuel}[€/L]}{10^6[€/M€]} \cdot \frac{V_{unloaded\ from\ the\ vessel}[m^3]}{V_{truck}[m^3] \cdot N^\circ_{days\ of\ prod.\ to\ store}[d]} \cdot \frac{8000[h/y]}{24[h/d]}$$

$$(2.23)$$

where d is the distance between the port and the end user (100 km). The cost of CO_2 emissions is calculated considering that diesel has specific CO_2 emissions (e_{fuel}) equal to 10.19 kg_{CO2}/gallon$_{diesel}$:

$$OPEX_{CO_2\ emissions}\left[\frac{M€}{y}\right] = E_{fuel}\left[\frac{kg_{CO_2}}{km}\right] \cdot d[km] \cdot 2 \cdot \frac{C_{CO_2\ emissions}[M€/t]}{10^3[kg/t]} \cdot \frac{V_{unloaded\ from\ the\ vessel}[m^3]}{V_{truck}[m^3] \cdot N^\circ_{days\ of\ prod.\ to\ store}[d]} \cdot \frac{8000[h/y]}{24[h/d]}$$

$$(2.24)$$

where E_{fuel} is calculated as:

$$E_{fuel}[kg_{CO_2}/km] = \frac{Cons_{fuel}[L_{diesel}/km] \cdot e_{fuel}[kg_{CO_2}/gallon_{diesel}] \cdot 264.2[gallon/m^3]}{1000[L/m^3]}. \quad (2.25)$$

2.6 Design Bases

The design bases for the present study are the following:

- green H_2 (produced by 100 MW electrolysers located in Algeria) continuously available with a flow rate of 20000 Nm^3/h, at pressure $P = 20$ bar and temperature $T = 25\ °C$;
- transport of the considered carrier from North Africa to North Italy in cargo ships powered by traditional fuel. This distance (approximately 2500 km) is indicated as the base case; a sensitivity analysis is then carried out in the range 2500–10000 km;
- two different H_2 applications:
  - industrial application (H_2 valley);
  - mobility sector application (HRS: Hydrogen Refuelling Station);
- variable end-user size depending whether it is a hydrogen valley (all the delivered hydrogen) or HRS (500 kg_{H2}/day);
- end-user distance from the port of 100 km;
- the transport of hydrogen from the port to the end user must be carried out by road. It is not considered appropriate to foresee, at the moment, different infrastructures (pipelines or railways);
- end-user H_2 purity ≥ 99.9 mol% in the case of H_2 valley, while ≥ 99.97 mol% in the case of HRS (as per the ISO 14687:2019 Hydrogen fuel quality—Product specification standard);
- H_2 end-user pressure equal to 30 bar in the case of H_2 valley and equal to 450/900 bar for HRS (two different pressure levels for refuelling buses/trucks and cars operating at 350 and 700 bar, respectively). In the following, the pressure of 900 bar is considered, as this is not significant for the comparison between the different chains.

Each hydrogen value chain can therefore be represented as in Fig. 2.1.

Each of the analysed chains involves: the conversion of hydrogen to the carrier considered at the departure terminal, the storage of the carrier at the departure terminal, the maritime transport from the departure terminal to the arrival terminal, the reconversion of the carrier to hydrogen at the arrival terminal and distribution of the hydrogen produced to the end users.

Fig. 2.1 Schematization of the value chains for hydrogen transport, considering the design bases introduced

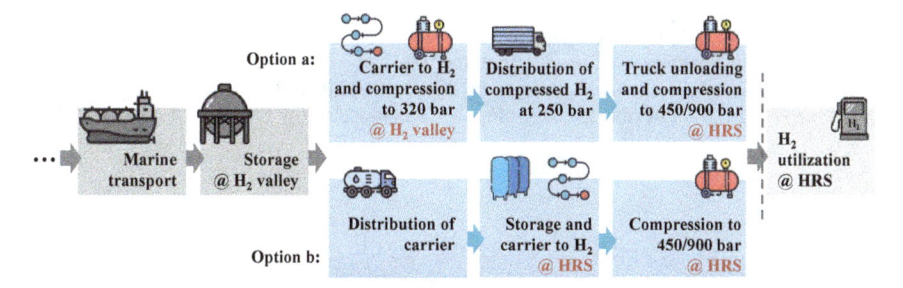

Fig. 2.2 Configurations available for reconversion and distribution operations in the HRS case: **a** centralized reconversion at the port and transport of compressed gaseous hydrogen; **b** distribution of liquid carrier and decentralized reconversion to the end users

As regards the destination of the hydrogen produced to the H$_2$ valley, the reconversion of the considered carrier is placed at the valley itself, to avoid transporting compressed hydrogen (and, consequently, the high compression costs) from the port of arrival to the valley.

However, as regards the destination of the hydrogen produced at the HRS, and therefore to various small end users, it is necessary to evaluate whether it is more cost-effective to reconvert the analysed carrier to gaseous hydrogen at the port of arrival (centralized reconversion) or at the end users (decentralized reconversion). The two configurations are shown in Fig. 2.2.

In option a, it is assumed to reconvert the carrier at the port. The hydrogen thus produced must be compressed to the pressure level necessary for its transport by truck to the end user (approximately 250 bar). In option b, however, it is assumed that the carrier is transported to the end user, where its conversion to hydrogen takes place.

These alternatives are compared in the case of liquefied hydrogen and ammonia carriers. In the case of LOHCs, however, the reconversion at the arrival terminal is defined *a priori* because it is considered more practical. In fact, in the case of reconversion of the carrier at the end users it would be necessary to provide, for each user, one or more dedicated trucks for transporting the dehydrogenated carrier to the port of arrival, to then allow it to be recirculated to the departure terminal.

References

1. Turton R, Bailie RC, Whiting WB, Shaeiwitz JA, Bhattacharyya D. Analysis, synthesis, and design of chemical processes, 4th ed. Pearson College Div.;2012.
2. www.chemengonline.com.
3. Okunlola A, Giwa T, Di Lullo G, Davis M, Gemechu E, Kumar A. Techno-economic assessment of low-carbon hydrogen export from Western Canada to Eastern Canada, the USA, the Asia-Pacific, and Europe. Int J Hydrogen Energy. 2022;47:6453–77.

4. Simbeck D, Chang E. Hydrogen supply: cost estimate for hydrogen pathways–scoping analysis, vol. 71. National Renewable Energy Laboratory;2002.
5. Reuß M, Grube T, Robinius M, Preuster P, Wasserscheid P, Stolten D. Seasonal storage and alternative carriers: a flexible hydrogen supply chain model. Appl Energy. 2017;200:290–302.

Chapter 3
Liquefied H$_2$ as Green H$_2$ Carrier

Abstract A techno-economic assessment is conducted on the liquefied hydrogen (LH$_2$) value chain, which includes hydrogen liquefaction, storage, maritime transport, distribution, regasification and compression. LH$_2$ is transported at -252 °C and atmospheric pressure. The cost-drivers of the value chain are the hydrogen liquefaction and regasification processes. Therefore, these processes are simulated using Aspen Plus V11® process simulator, in order to estimate their capital and operating expenditures from the obtained material and energy balances.

Keywords Liquefied hydrogen · Economic assessment · Cryogenic equipment · Value chain · Hydrogen liquefaction · Hydrogen storage

3.1 Introduction

The first considered hydrogen carrier is hydrogen itself in its liquid state (LH$_2$). Under saturated liquid conditions at 1.3 bar ($T = -252$ °C), hydrogen has a density of approximately 70 kg/m^3, allowing it to be transported in properly insulated tanks.

The value chain for liquid hydrogen transportation, as shown in Fig. 3.1, includes the following steps:

- *Liquefaction*. Liquefaction is the physical transformation process of H$_2$ from the gaseous state at ambient temperature (25 °C) to the liquid state at -252 °C and 1.3 bar pressure.
- *Storage at the departure terminal*. Storage takes place in properly insulated spherical tanks to limit the heat exchange with the surrounding environment. The boil-off gas is recycled to the liquefaction process.
- *Maritime transport*. Transport occurs on ships equipped with insulated spherical tanks. The boil-off gas is flared.
- *Storage at the arrival terminal*. Storage takes place into properly insulated spherical tanks. The amount of boil-off gas is considered negligible for port storage. During storage at the utilization site, the boil-off gas is sent to the regasification process.

Fig. 3.1 Liquefied hydrogen value chain

- *Regasification.* Regasification is the physical transformation process of H_2 from the liquid state at -252 °C and 1.3 bar pressure to the gaseous state at ambient temperature (25 °C) and user's pressure.
- *Distribution.* Distribution can occur upstream or downstream of the regasification process. In the first case, liquid hydrogen would be distributed via tanker trucks and regasified on site. In the second case, gaseous hydrogen would be distributed, having been compressed to increase its density and stored in appropriate cylindrical tanks (tube trailers), via trailer trucks.

3.2 Hydrogen Liquefaction

Various configurations for the hydrogen liquefaction process have been explored in literature. Among these, the process described by Cardella et al. [1], which utilizes a mixed refrigerant (MR) Joule–Thomson (J-T) cycle for precooling and a high-pressure hydrogen Claude cycle for cryogenic cooling, has been identified as the most balanced in terms of complexity, related to *CAPEX*, and specific electricity consumption (SEC), related to *OPEX*, considering the relatively small-capacity plant under consideration (43.2 tpd) [2]. Therefore, this process is selected for simulation in Aspen Plus V11® with the aim to obtain material and energy balances, necessary for the economic assessment. The Peng-Robinson thermodynamic package, as modified by Restelli et al. [3] to represent the behaviour of equilibrium-hydrogen, the temperature-dependent equilibrium mixture of ortho- and para-hydrogen, is chosen for this simulation. This choice is justified by the model's combined accuracy and ease of implementation within the simulator. An isentropic efficiency of 0.8 is assumed for compressors and turbine expanders. The Process Flow Diagram (PFD) is depicted in Fig. 3.2.

Referring to Fig. 3.2, the fed hydrogen (GH_2) undergoes precooling to -173 °C in multi-stream plate-fin heat exchanger HX-100, where the precooling MR flows counter-currently. The composition of this mixture has been optimized by Cardella et al. [1] to reduce temperature differences between the cold and hot composite curves inside HX-100 and consists of 14% nitrogen, 30% methane, 31% ethane and 25% iso-butane on a molar basis. The advantage of refrigerant mixtures is that they evaporate and condense over a wider temperature range than pure refrigerants. The MR is compressed to 50 bar with inter-refrigeration using service water. Downstream of the compressor C-104 post-refrigerator, the high-pressure stream 19 is separated

3.2 Hydrogen Liquefaction

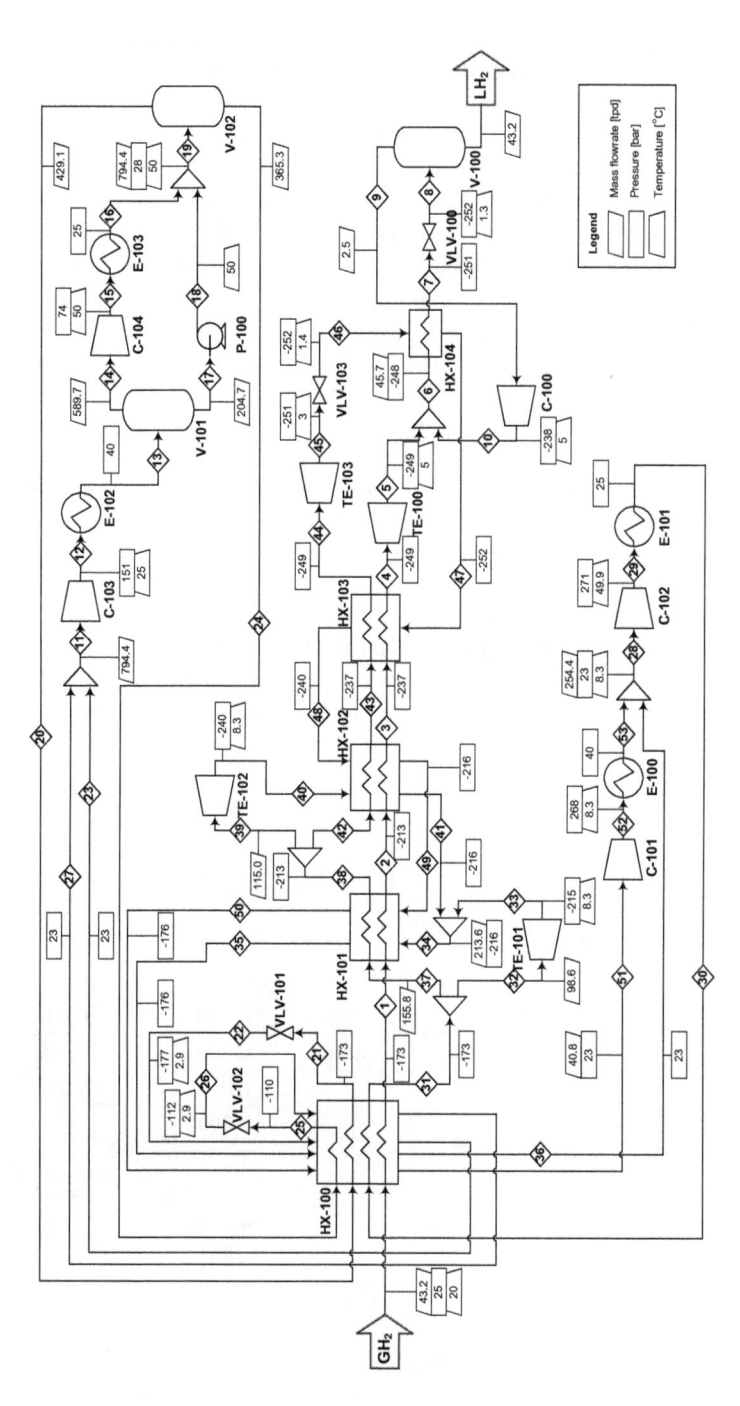

Fig. 3.2 PFD of the hydrogen liquefaction process

into its vapour phase 20 and liquid phase 24. The 20 and 24 streams are pre-cooled in HX-100 and expanded to 2.9 bar in J-T valves, VLV-101 and VLV-102, to reach two different precooling temperatures: $-177\ °C$ and $-112\ °C$, respectively. The low-pressure streams 22 and 26 act as cold fluids in the HX-100 exchanger and are mixed at its outlet. The precooling section described above is located inside a precooling coldbox, to limit the heat exchange with the surrounding environment.

The design of the high-pressure H_2 Claude cycle has been optimized to reduce temperature differences in the plate-fin heat exchangers HX-101 to HX-104. This cycle is a modification from the basic Claude cycle design. On the cold side of the heat exchangers in the cryogenic cooling cold box, the J-T part of the Claude cycle is designed with a turbine (TE-103) upstream of the final J-T valve (VLV-103), expanding to 1.4 bar. The Brayton part of the Claude cycle is designed with two turbines TE-100 and TE-101 expanding to the intermediate pressure of 8.3 bar to remove heat from the processed hydrogen stream at two different temperature levels: $-215\ °C$ and $-240\ °C$ at the outlet of TE-101 and TE-102, respectively. At the outlet of the last heat exchanger (HX-104), the processed hydrogen is expanded in J-T valve VLV-100 to the storage pressure of 1.3 bar, with the aim of cooling and partially liquefying the H_2 stream. The phase separator V-100 is used to split the vapour and liquid phases of the depressurized stream 8. The liquid stream (LH_2) is sent to the storage tank, while the vapour stream 9 is recompressed and recycled at a tie-in point upstream the heat exchanger HX-104. The cryogenic section described above is located inside a cryogenic cold box. The energy balance of the hydrogen liquefaction process is reported in Table 3.1 in terms of thermal duty ($Q > 0$ supplied to the system) and electric power ($W > 0$ supplied to the system).

Table 3.1 Energy balance of the process in Fig. 3.2

Thermal energy		Electric energy	
Equipment	Q [kW]	Equipment	W [kW]
E-100	−1538.5	C-100	4.7
E-101	−10639.6	C-101	1665.3
E-102	−2184.3	C-102	10661.6
E-103	−2110.0	C-103	2094.5
		C-104	543.0
		P-100	1.2
		TE-100	−9.4
		TE-101	−450.3
		TE-102	−236.3
		TE-103	−27.3

3.3 Liquefied H$_2$ Regasification

Liquefied hydrogen regasification is the process of converting liquid hydrogen back into its gaseous state. It is a relatively simple process, involving the pumping of liquefied hydrogen and its heating to ambient temperature using service water available at 30 °C. Regasification of the transported hydrogen can occur at a single dedicated hub (centralized regasification) or in multiple hubs (decentralized regasification), typically located next to the utilization sites. When delivering hydrogen to a H$_2$ valley, it is convenient to conduct regasification at a hub located nearby. However, when hydrogen refuelling stations are the destination sites, it is necessary to investigate whether it is more cost-effective to perform centralized regasification at the unloading port, involving downstream distribution of compressed gaseous hydrogen to the HRS via trucks, or to conduct decentralized regasification at the refuelling stations, involving upstream distribution of liquefied hydrogen.

3.3.1 Centralized Regasification

The PFD of the centralized regasification process at the H$_2$ valley is depicted in Fig. 3.3. The process is simulated in Aspen Plus V11$^®$. Referring to Fig. 3.3, the pump discharge pressure is set to 30 bar to meet the needs of the industries located in the vicinity of the valley. For all the cases, it is assumed a pump efficiency of 0.6 and service water as hot fluid in the vaporizer. The energy balance of the centralized regasification process at the H$_2$ valley is reported in Table 3.2.

In case HRS are the end users of the delivered hydrogen, if centralized regasification is performed at the port of arrival, pumping to 320 bar is necessary to feed the tube trailers, operating at 250 bar, for compressed hydrogen distribution.

Fig. 3.3 PFD of the centralized liquefied hydrogen regasification process at the H$_2$ valley

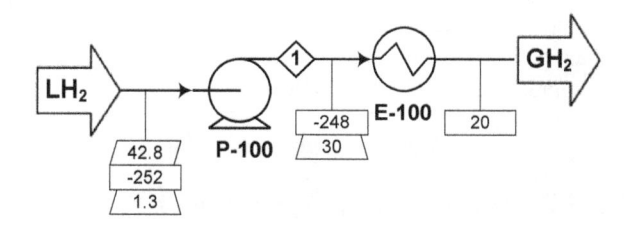

Table 3.2 Energy balance of the process in Fig. 3.3

Thermal energy		Electric energy	
Equipment	Q [kW]	Equipment	W [kW]
E-100	2102.6	P-100	33.6

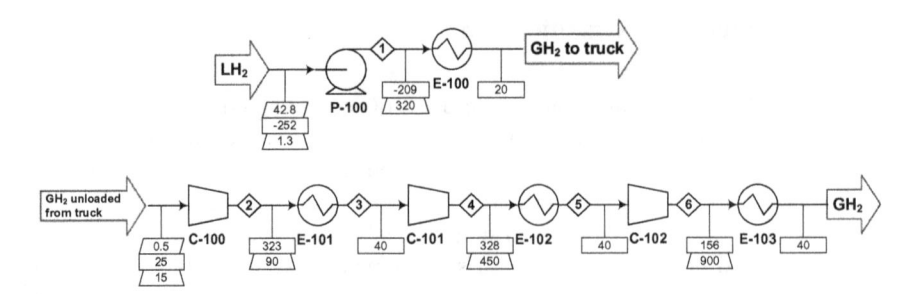

Fig. 3.4 PFD of the centralized liquefied hydrogen regasification process at the unloading port and decentralized hydrogen compression process at the HRS

Table 3.3 Energy balance of the process in Fig. 3.4

Thermal energy		Electric energy	
Equipment	Q [kW]	Equipment	W [kW]
E-100	1842.0	P-100	373.1
E-101	-2039.7 ($N_{HRS} \bullet Q_{E\text{-}101}$)	C-100	2200.6 ($N_{HRS} \bullet W_{C\text{-}100}$)
E-102	2112.5 ($N_{HRS} \bullet Q_{E\text{-}102}$)	C-101	2226.9 ($N_{HRS} \bullet W_{C\text{-}101}$)
E-103	858.0 ($N_{HRS} \bullet Q_{E\text{-}103}$)	C-102	1025.6 ($N_{HRS} \bullet W_{C\text{-}102}$)

In case of hydrogen delivery to HRS with centralized regasification at the port of arrival and compressed gaseous hydrogen distribution, compression to 900 bar at the refuelling stations is required after tube trailer discharge. An efficiency of 0.65 is assumed for the refuelling station compressors. The PFD of the centralized regasification process at the unloading port, together with the compression process occurring at each refuelling station, is represented in Fig. 3.4. Table 3.3 details the energy balance of the centralized regasification process at the unloading port and the compression process at the HRS.

3.3.2 Decentralized Regasification

In the case of regasification at the HRS end users, whose PFD is shown in Fig. 3.5, the processed flow rate is that required by each individual end user, i.e. 500 kg/d, and the number of processes is equal to the number of users that the total flow rate can satisfy (approximately 86). At each individual user, LH$_2$ is pumped to 900 bar, avoiding the need for downstream gas compression, and heated. In Table 3.4, the energy balance of the decentralized regasification process at the HRS is reported.

Fig. 3.5 PFD of the decentralized liquefied hydrogen regasification process at the HRS

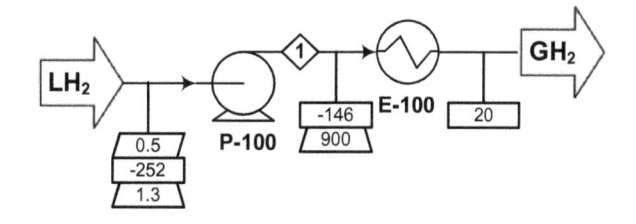

Table 3.4 Energy balance of the process in Fig. 3.5

Thermal energy		Electric energy	
Equipment	Q [kW]	Equipment	W [kW]
E-100	1383.2 ($N_{HRS} \bullet Q_{E-100}$)	P-100	1062.2 ($N_{HRS} \bullet W_{P-100}$)

3.4 Techno-Economic Assessment

The Block Flow Diagram (BFD) of the liquefied hydrogen value chain is reported in Fig. 3.6. In both cases, whether delivering hydrogen to a H_2 valley or to hydrogen refuelling stations, losses of hydrogen occur due to boil-off during sea transport. When hydrogen is destined for HRS, it can serve approximately 86 stations.

Following the methodology outlined in Chap. 2, the capital and operating costs of the liquefaction process (Sect. 3.2), the centralized regasification process (Sect. 3.3.1) and the decentralized regasification process (Sect. 3.3.2) are determined.

In particular, for liquefaction, the *CAPEX* are 115.85 M€ and the total bare module cost of equipment can be subdivided based on the equipment category, as shown in Fig. 3.7. Compressor costs are the most significant cost item, followed by the cost of heat exchangers.

Figure 3.8 juxtaposes the result obtained in this study (red bar) with cost data for the hydrogen liquefaction process from literature, categorized by process type. To ensure a meaningful comparison among different estimates, the literature values are adapted to account for hydrogen liquefaction capacity using the six-tenths rule and are normalized to the base year using the CEPCI. Figure 3.8 allows to notice the significant variability in literature estimates, particularly concerning the liquid nitrogen (LN_2) precooled Claude cycle. Notably, the Nexant [4] value demonstrates a considerable overestimation compared to others. The value derived from this study falls between the estimations for the MR cascade process and the LN_2 precooled process, which is reasonable given the intermediate complexity level.

The *OPEX* amount to 101.31 M€/y in the present scenario and are primarily influenced by the high electricity cost.

For centralized regasification in the H_2 valley case, the *CAPEX* are 2.08 M€, with *OPEX* at 4.14 M€/y. In the HRS case, centralized regasification at the unloading

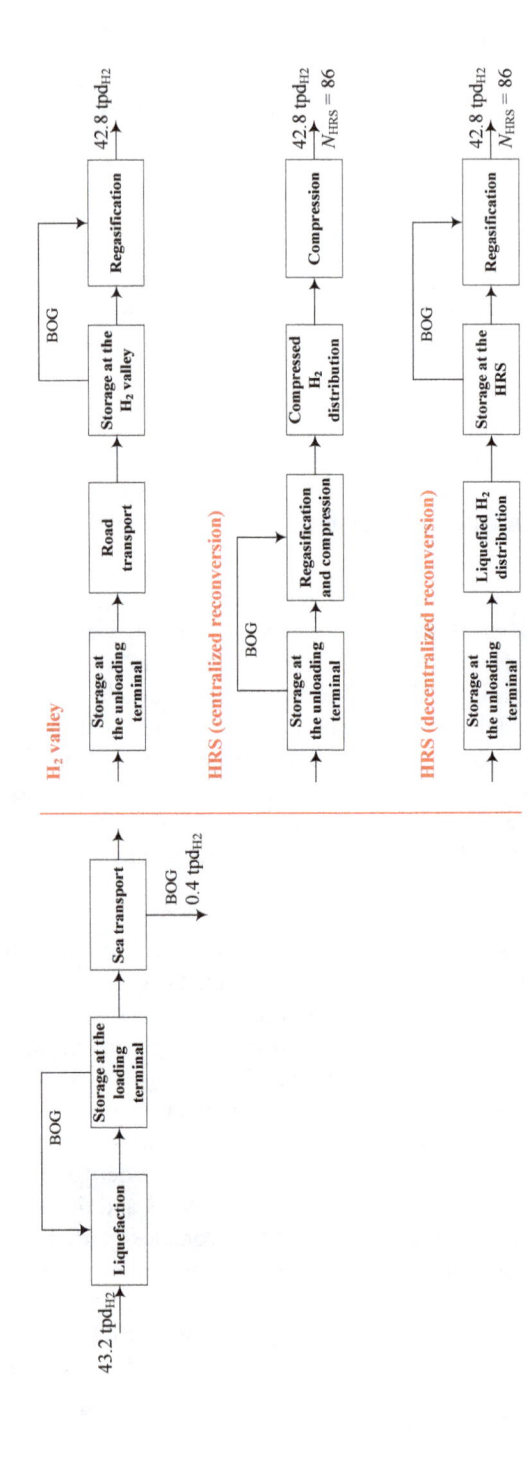

Fig. 3.6 BFD of the LH$_2$ value chain for the base case ($D = 2500$ km)

3.4 Techno-Economic Assessment

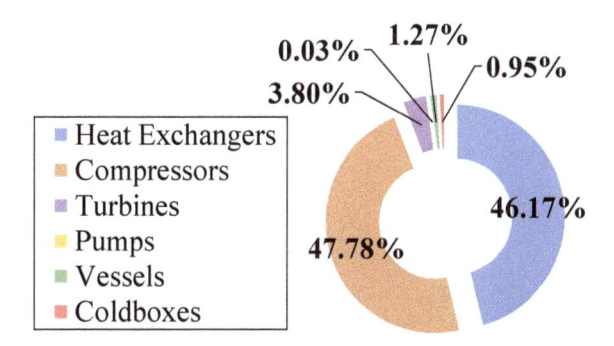

Fig. 3.7 Breakdown of the bare module cost of equipment for the hydrogen liquefaction process

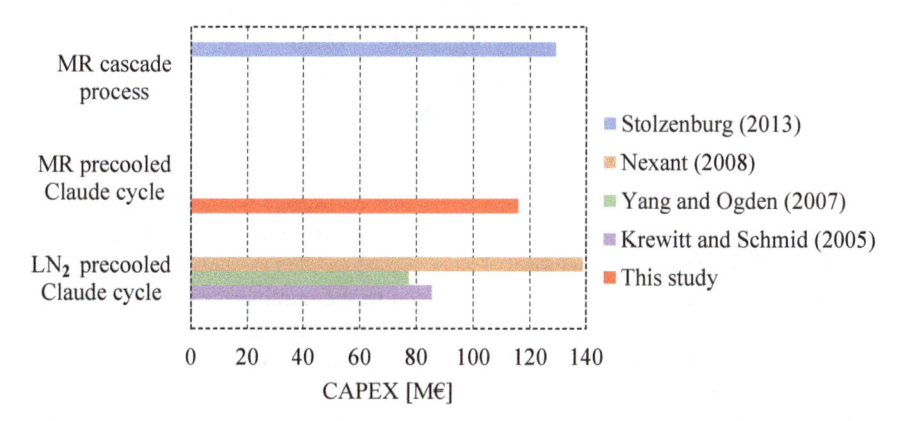

Fig. 3.8 *CAPEX* [M€] of the hydrogen liquefaction process: comparison between the results obtained in this study and the literature [4–7]

port incurs *CAPEX* and *OPEX* of 0.40 M€ and 1.20 M€/y, respectively. In addition, costs for compression at the refuelling stations must be considered, amounting to 1.25 M€ for *CAPEX* and 1.16 M€/y for *OPEX* for each end user. For decentralized regasification in the HRS case, the *CAPEX* and *OPEX* are 0.86 M€ and 0.81 M€/y, respectively, for each end user with no further compression required in this configuration.

Regarding the sea transport of LH$_2$, liquefied natural gas (LNG) transport can be assumed as a reference, given the lack of data on liquefied hydrogen carriers. Table 3.5 presents the annual reference year and *CAPEX* for LNG ships of different capacities. These costs are reported to the base year using the CEPCI and interpolated with a power law (Eq. 3.1).

$$CAPEX_{vessel} = 0.7316 \cdot V_{vessel}^{0.496} \qquad (3.1)$$

Table 3.5 Reference year and CAPEX for LNG vessels of different capacities [8]

V_{vessel} [m^3]	Reference year	Ref. *CAPEX* [M$]
6000	2015	50
7500	2018	37
12000	2014	50
28000	2018	80
30000	2014	105

Table 3.6 Power and crew size for LNG vessels of different capacities

Reference	Volume [m^3]	Power [kW]	Crew [people]
LNG vessel data sheet[1]	7500	3000	16
LNG vessel data sheet[2]	15000	4500	20
LNG vessel data sheet[3]	10000	3360	18
LNG vessel data sheet[4]	30000		24
LNG vessel data sheet[5]	40000	6000	

Table 3.6 reports the power and crew size derived from the technical data sheets of LNG ships of different capacities.

The *CAPEX* for LH$_2$ sea transport is related to the purchase of the ship and are derived, for the capacity corresponding to different harbour-to-harbour distances. Operating costs are related to labour costs, and hence to the crew size, which is determined from the technical data sheet of a LNG vessel with almost the same capacity as that considered, fuel consumption, which is function of the ship's power, CO$_2$ emissions, maintenance and insurance. Table 3.7 summarizes *CAPEX* and *OPEX* for the different cases under consideration.

Throughout sea transport, approximately 0.2% of hydrogen is lost daily due to the boil-off phenomenon. While this released hydrogen gas could potentially be utilized for onboard heating or power generation, for the purposes of this analysis, it is presumed that its value is not recovered, and instead, it is directed to a flare.

Storage is carried out at approximately −252 °C and at a pressure slightly higher than atmospheric pressure, 1.3 bar. Liquefied hydrogen is stored in spherical tanks (minimum surface-to-volume ratio) equipped with vacuum perlite insulation to limit

---

[1] https://products.damen.com/-/media/products/images/clusters-groups/shipping/liquefied-gas-carrier/liquefied-gas-carrier-7500-lng/downloads/product_sheet_damen_liquefied_gas_carrier_7500_lng_10_2017.pdf?rev=7e8230a833e1458ca06c29a4b7a6b5e4.

[2] https://mes.it/15000-cbm-lng-carrier-bunkering/

[3] https://www.wartsila.com/docs/default-source/product-files/sd/merchant/lng/data-sheet-ship-design-lng-bunker-wsd59-10k.pdf?sfvrsn=b94cc545_8.

[4] https://cdn.wartsila.com/docs/default-source/product-files/sd/merchant/lng/wsd50-30k-lng-carrier-ship-design-O-data-sheet.pdf?sfvrsn=e8b38445_8.

[5] https://sagalng.com/wp-content/uploads/2017/05/40k-Wuhu-max-Data-sheet.pdf.

3.4 Techno-Economic Assessment

Table 3.7 *CAPEX* [M€] and *OPEX* [M€/y] for sea transport in the LH$_2$ value chain, depending on the distance *D* travelled by ship

D [km]	V$_{vessel}$ [m³]	CAPEX [M€]	OPEX [M€/y]
2500 (base)	6700	57.81	9.69
5000	11400	75.25	12.63
10000	20700	101.16	16.95

the boil-off losses. Similar to LNG storage, a certain amount of liquid must remain in the tank to keep it cold, approximately 4% of the tank volume.

A power law for estimating the *CAPEX* of LH$_2$ tanks is obtained by interpolating the cost data reported in Nexant Inc. [4], after being inflation-adjusted to the base year and multiplied by a factor of 1.3 to account for the installation cost.

$$CAPEX_{tank} = 0.057 \cdot V_{tank}^{0.6891} \tag{3.2}$$

For the storage of liquefied hydrogen at the full-capacity terminals, fixed costs are related to the purchase of the tank and are derived, for the capacity corresponding to different harbour-to-harbour distances. *OPEX* are related to maintenance and insurance.

For the case involving the transport of hydrogen to a H$_2$ valley, the purchase of 3 full-capacity storage tanks is required: one at the loading harbour, one at the unloading harbour and one at the H$_2$ valley. For the case involving hydrogen delivery to HRS, 2 full-capacity tanks are required at the harbours. In addition, if regasification is carried out at the HRS, a small-capacity LH$_2$ tank (9 m³) is necessary to feed the decentralized regasification process at each refuelling station, with *CAPEX* of 0.088 M€ (Nexant Inc. [4], cost adjusted for inflation in 2022).

Table 3.8 reports the *CAPEX* of one full-capacity storage tank for the different cases under consideration.

Distribution can take place either upstream or downstream of the regasification process. In the former case, LH$_2$ would be distributed using trucks, designed to operate at −252 °C and 1.3 bar, with the regasification process carried out at the destination site. In the latter case, gaseous H$_2$ would be distributed using trucks equipped with suitable cylindrical tanks (referred to as tube trailers) designed for operation at ambient temperature and 250 bar. The hydrogen gas is released from the tube trailers by reducing the pressure to 15 bar.

Table 3.8 *CAPEX* [M€] and *OPEX* [M€/y] for one liquefied hydrogen full-capacity storage tank in the LH$_2$ value chain, depending on the distance *D* travelled by ship

D [km]	V$_{tank}$ [m³]	CAPEX [M€]	OPEX [M€/y]
2500 (base)	7300	26.19	2.62
5000	12500	37.94	3.79
10000	22700	57.23	5.72

Table 3.9 CAPEX [M€] and OPEX [M€/y] for liquefied hydrogen distribution in the LH$_2$ value chain, depending on the distance D travelled by ship

D [km]	n_{trucks}	CAPEX [M€]	OPEX [M€/y]
2500 (base)	6	8.93	1.59
5000	6	8.93	1.59
10000	6	8.93	1.59

Table 3.10 CAPEX [M€] and OPEX [M€/y] for compressed hydrogen distribution in the LH$_2$ value chain, depending on the distance D travelled by ship

D [km]	n_{trucks}	CAPEX [M€]	OPEX [M€/y]
2500 (base)	43	37.64	11.77
5000	43	37.64	11.77
10000	42	36.76	11.50

In the H$_2$ valley case and in the HRS case with decentralized regasification at the end user, liquefied hydrogen is distributed. A LH$_2$ trailer with a net capacity of 4000 kg is selected, with CAPEX of 1.19 M€ (DOE [9], cost adjusted for inflation in 2022). Based on the assumptions described in Chap. 2, it is necessary to purchase 6 trailer trucks. Operating costs are related to labour costs, fuel consumption, and CO$_2$ emissions. Table 3.9 summarizes CAPEX and OPEX for the different case studies.

If in the HRS case, regasification is carried out at the destination port, compressed gaseous hydrogen is distributed and a compressed gaseous H$_2$ trailer with a net capacity of 500 kg and CAPEX of 0.57 M€ (DOE [9], cost adjusted for inflation in 2022) is adopted.

According to the assumptions outlined in Chap. 2, it is necessary to purchase approximately 43 trailer trucks. Table 3.10 summarizes CAPEX and OPEX for the different cases under consideration.

References

1. Cardella U, Decker L, Sundberg J, Klein H. Process optimization for large-scale hydrogen liquefaction. Int J Hydrogen Energy. 2017;42:12339–54.
2. Restelli F, Spatolisano E, Pellegrini LA, Cattaneo S, de Angelis AR, Lainati A, et al. Liquefied hydrogen value chain: a detailed techno-economic evaluation for its application in the industrial and mobility sectors. Int J Hydrogen Energy. 2024;52:454–66.
3. Restelli F, Spatolisano E, Pellegrini LA. Hydrogen liquefaction: a systematic approach to its thermodynamic modeling. Chem Eng Trans. 2023;99:433–8.
4. Nexant Inc., Air Liquide, Argonne National Laboratory, Chevron Technology Venture, Gas Technology Institute, National Renewable Energy Laboratory, et al. H2A hydrogen delivery infrastructure analysis models and conventional pathway options analysis results;2008.
5. Stolzenburg K, Mubbala R. Hydrogen liquefaction report. Integrated design for demonstration of efficient liquefaction of hydrogen (IDEALHY), FCH JU;2013.
6. Yang C, Ogden J. Determining the lowest-cost hydrogen delivery mode. Int J Hydrogen Energy. 2007;32:268–86.

References

7. Krewitt W, Schmid S. WP 1.5 common information database D 1.1 fuel cell technologies and hydrogen production/distribution options;2005.
8. Fikri M, Hendrarsakti J, Sambodho K, Felayati F, Octaviani N, Giranza M, et al. Estimating capital cost of small scale LNG carrier. p. 5–6.
9. DOE Multi-year research, development and demonstration plan—3.2 hydrogen delivery;2015.

Chapter 4
Ammonia as Green H$_2$ Carrier

Abstract Ammonia (NH$_3$) holds promise for decarbonization of both energy and industrial sectors. Thanks to its high H$_2$ density, it is also a good candidate as H$_2$ carrier. A techno-economic assessment is conducted on the liquefied ammonia value chain, which includes ammonia synthesis, storage, maritime transport, distribution, ammonia cracking and compression. Liquefied NH$_3$ is transported at −30 °C and atmospheric pressure. The cost-drivers of the value chain are the ammonia synthesis and cracking processes. Therefore, these processes are simulated using Aspen Plus V11® process simulator, in order to estimate their capital and operating expenditures from the obtained material and energy balances.

Keywords Green ammonia · Economic assessment · Value Chain · Ammonia cracking · Hydrogen carrier

4.1 Introduction

Ammonia (NH$_3$) is one of the main building blocks of Chemical Engineering, normally synthesized from natural gas and mainly used as a raw material in the production of fertilizers. However, it can also be used to store clean hydrogen, transporting it in refrigerated tanks and splitting it into its components, N$_2$ and H$_2$, through an endothermic cracking process, once it reaches its destination. The resulting mixture is then purified, and the produced hydrogen distributed. Figure 4.1 shows the value chain of ammonia as an energy carrier for hydrogen. This value chain includes:

- *Ammonia synthesis.* NH$_3$ production involves a chemical process that combines N$_2$, obtained by air fractionation, and H$_2$, produced through green electricity-driven electrolysers, to form NH$_3$. Cooling the reaction products serves to recover the reagents and obtain liquefaction at a pressure of 1.3.
- *Storage at the departure terminal.* Storage takes place in properly insulated spherical tanks. The boil-off gas is reliquefied.

Fig. 4.1 Liquefied ammonia value chain

- *Maritime transport.* Transport takes place on ships equipped with insulated spherical tanks. The boil-off gas is flared.
- *Storage at the arrival terminal.* Storage takes place in properly insulated spherical tanks. The amount of boil-off gas is assumed to be negligible for port storage. During storage at the utilization site, the boil-off gas is sent to the cracking process.
- *Ammonia cracking.* NH_3 cracking is a chemical process where NH_3 is decomposed into N_2 and H_2. Downstream hydrogen compression is eventually required to meet user's pressure.
- *Distribution.* Distribution can take place upstream or downstream of the cracking process. In the first case, liquid ammonia would be distributed via tanker trucks and cracking would take place on-site. In the second case, gaseous hydrogen would be distributed, after being compressed, using tube trailer trucks.

The main cost elements of the chain in Fig. 4.1 are the synthesis and cracking of ammonia, detailed in Sects. 4.2 and 4.3.

4.2 Ammonia Synthesis

To carry out the synthesis of NH_3 and, ultimately, estimate the fixed and operating costs relevant to the ammonia chain, it is necessary to provide a nitrogen supply to the process. The production of nitrogen from air is therefore considered, rather than introducing a complete air separation unit (ASU), as the process under consideration does not foresee a possible destination for the oxygen produced. In the literature, several schemes aimed at producing nitrogen from air are available. In particular, the Air Products scheme proposed by Agrawal and Thorogood [1] is considered, whose simulation in Aspen Plus V11® is represented in Fig. 4.2. The process involves the separation of the supplied air, compressed up to 8 bar and pre-cooled to its dew point in the main exchanger HX-106, via cryogenic distillation. The supplied air is then sent to the rectification section (column T-100), operating at high pressure ($P = 8$ bar) and consisting of 50 theoretical trays. Here, it is separated into a vapour phase top product, made up of pure nitrogen (residual O_2 content of 2.14 ppm), and a bottom product, consisting of an approximately equimolar mixture of oxygen and nitrogen. This mixture, after expansion to approximately 3 bar, is fed to the stripping section (column T-101), which separates at the top a stream whose composition is close to that of air and at the bottom air enriched in oxygen (O_2 content of 57.57 mol%).

4.2 Ammonia Synthesis

The latter stream, after being sent to the process-process exchangers HX-107 and HX-106, constitutes one of the products of the proposed scheme (stream O_2-rich in Fig. 4.2). The cooling duty necessary to condense part of the top product of the high-pressure section, which constitutes the reflux of the rectification section, is provided by the evaporation of the bottom product of the low-pressure section. The approach temperature difference in HX-107 is 1 °C. The process is autothermal: the energy demands of the separation are concentrated in the cost of compressing the supplied air. The nitrogen stream thus produced (stream 71 in Fig. 4.2) constitutes the feed for the ammonia synthesis stage.

The Haber-Bosch ammonia synthesis process, widely consolidated and used on a large scale, is based on reaction 4.1 and is usually catalysed by iron-based catalysts.

$$N_2 + 3H_2 \rightarrow 2NH_3 \qquad (4.1)$$

To simulate the ammonia synthesis section in Aspen Plus V11® and, therefore, estimate its costs, reference is made to the template available in the process simulator, modifying it significantly both to remove the steam reforming production unit and to adapt it to a different capacity (different sizes of reaction modules). The performances of the kinetic model implemented in the simulator [2] and the specified operating conditions are appropriately validated and adjusted. The reaction section, consisting of 3 adiabatic beds with intermediate cooling, is scaled using the procedure reported in [3]. The nitrogen feed (stream 71 in Fig. 4.2), coming out of the air separation process at approximately 29 °C and 8 bar, is compressed up to 20 bar and mixed with the hydrogen supplied (stream GH_2 in Fig. 4.2). The resulting mixture is compressed to 193 bar in a three-stage compressor with intercooling. Downstream of the last post-cooler E-102, the feed, at 40 °C and 193 bar, is mixed with the recirculation streams, the vapours exiting the V-100, V-101 and V-102 separators, suitably compressed. The mixture is compressed to 200 bar in C-103 and heated to 347 °C via a series of process-process heat exchangers (HX-100, HX-101, HX-102), which exploits the high enthalpy content of the streams exiting the reactor beds, before being sent to the first catalytic stage R-100.

The stream exiting the last reactor stage R-102 contains, in addition to ammonia, hydrogen and nitrogen which must be separated and recirculated. The objective is to obtain liquid ammonia at a pressure slightly higher than ambient one, with a purity higher than 99.95%. To achieve the required specifications, the stream is first cooled in HX-103 to 110 °C, providing heat to a steam Rankine cycle, then in the train of exchangers HX-100 and E-103.

The biphasic stream coming out of E-103 is separated in the V-100 separator: the vapour is recirculated to the reaction section, while the liquid is further purified in a series of downstream expansion and cooling units, using liquid ammonia as refrigerant fluid.

The energy balance of the ammonia synthesis process is reported in Table 4.1 in terms of thermal duty ($Q > 0$ supplied to the system) and electric power ($W > 0$ supplied to the system).

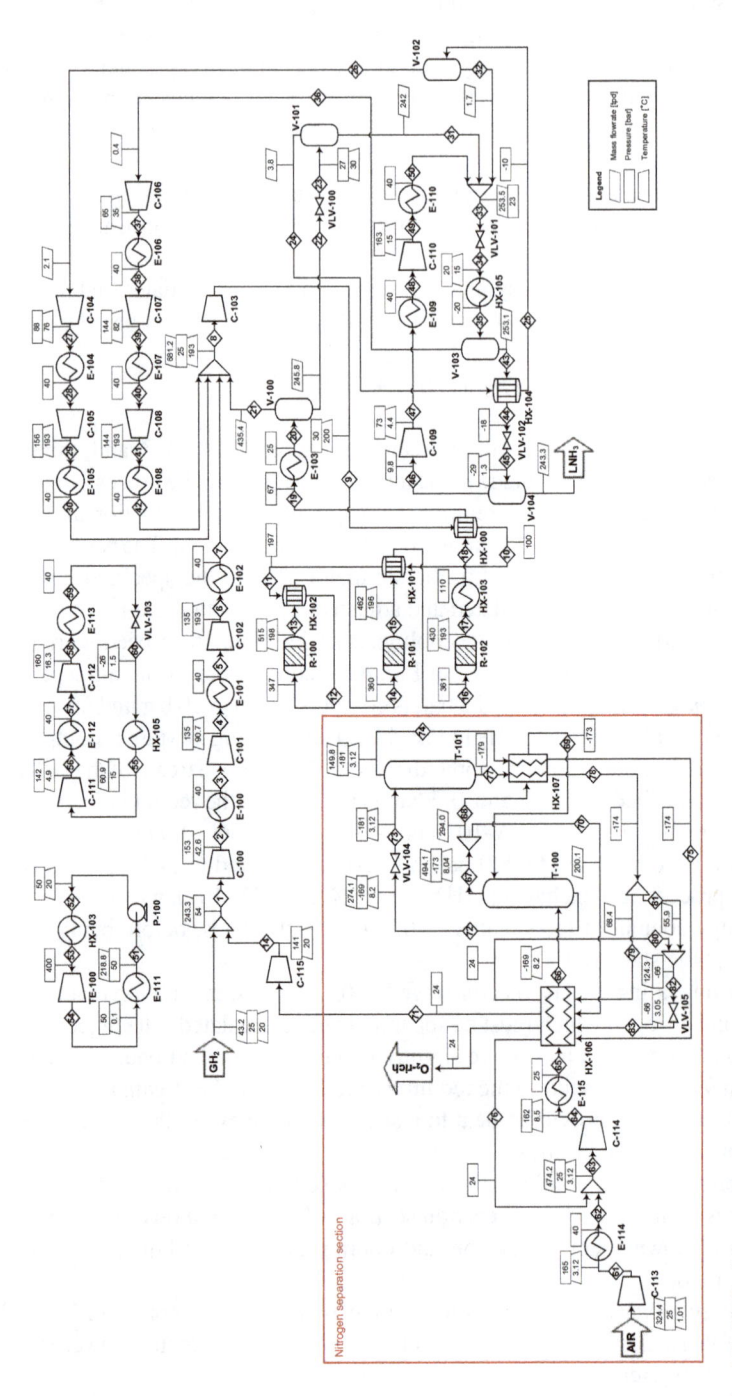

Fig. 4.2 PFD of the ammonia synthesis process

4.3 Ammonia Cracking

Table 4.1 Energy balance of the process in Fig. 4.2

Thermal energy		Electric energy	
Equipment	Q [kW]	Equipment	W [kW]
E-100	−1101.3	C-100	965.5
E-101	−935.0	C-101	942.1
E-102	-957.9	C-102	981.9
E-103	−3065.3	C-103	126.9
E-104	−3.7	C-104	7.3
E-105	−9.2	C-105	9.0
E-106	−0.3	C-106	1.0
E-107	−1.2	C-107	1.2
E-108	−1.4	C-108	1.2
E-109	−8.4	C-109	23.6
E-110	−33.9	C-110	29.6
E-111	−6069.1	C-111	174.9
E-112	−163.8	C-112	177.2
E-113	−988.5	C-113	534.9
E-114	−479.3	C-114	709.4
E-115	−771.9	C-115	272.0
		P-100	6.5
		TE-100	−2013.2

4.3 Ammonia Cracking

The dissociation of ammonia is a highly endothermic process, described by reaction 4.2, which generally occurs at high temperature, which represents the main challenge in using this technology on an industrial scale.

$$2NH_3 \rightarrow N_2 + 3H_2 \qquad (4.2)$$

As a result of the first studies and applications, and in parallel with the search for catalysts for thermal decomposition, alternative methods have been proposed in the literature to provide the activation energy necessary for the reaction. Among all the technologies reported in the literature, thermocatalytic cracking is the most mature from a technological point of view. In fact, on a commercial level, small ammonia cracking reactors are available (size between 1 Nm^3/h and 1000 Nm^3/h), which operate at temperatures between 800 and 1000 °C with nickel-based catalysts. The endothermicity of the reaction is supported by electricity. The process is mainly used in the steel industry, where it is necessary to recreate a reducing environment during metal processing.

In the following discussion, the process simulations developed for the cases of centralized and decentralized cracking are reported, assuming the thermocatalytic nickel-based technology.

4.3.1 Centralized Cracking

In the H_2 valley case, ammonia cracking occurs in a centralized mode; all ammonia stored at the arrival port is converted into hydrogen at the H_2 valley. The centralized cracking process, as simulated in Aspen Plus V11®, is shown in Fig. 4.3. The ammonia (stream LNH$_3$ in Fig. 4.3), upon reaching the H_2 valley, is pumped up to approximately 30 bar, preheated in heat exchanger HX-100 by exploiting the high enthalpy content of the reaction products, and then sent to the cracking reactor. The reactor R-100 is simulated with the RGibbs module of Aspen Plus: the conversion of ammonia is, therefore, that of thermodynamic equilibrium at the reactor's temperature and pressure, *i.e.*, 30 bar and 900 °C. This temperature is in line with the operating conditions of commercially available nickel-based catalysts. The heat duty necessary for the cracking reaction is supplied through combustion of part of the ammonia arriving at the H_2 valley, appropriately mixed with the waste streams with a high hydrogen content exiting the purification section. Inside the combustion furnace, the ammonia flow rate is regulated to generate the heat necessary for the cracking reaction, while the air flow rate entering the same unit is kept slightly above stoichiometric, to guarantee complete combustion.

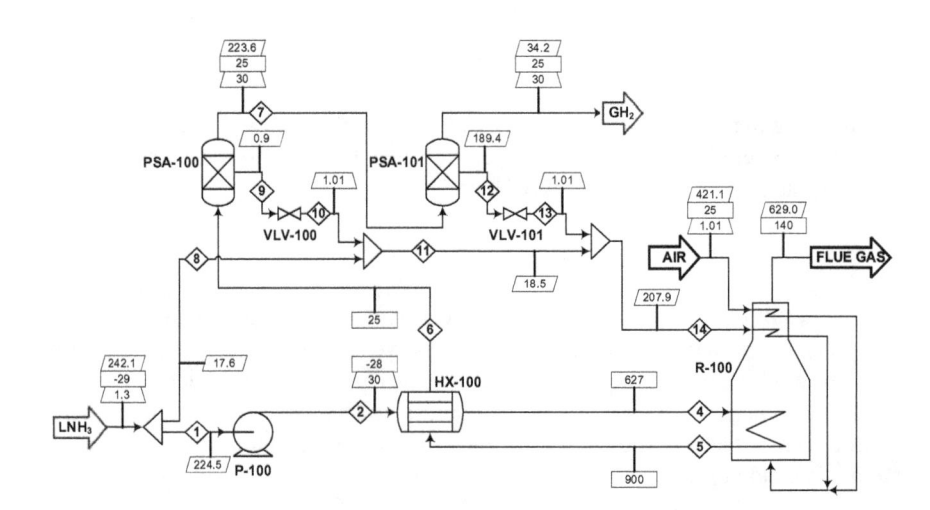

Fig. 4.3 PFD of the centralized ammonia cracking process at the H_2 valley

Table 4.2 Energy balance of the process in Fig. 4.3

Electric energy	
Equipment	W [kW]
P-100	23.7

Downstream of the reaction stage, the separation of the produced hydrogen from the unconverted ammonia and nitrogen takes place via pressure swing adsorption (PSA). The outlet hydrogen (stream GH_2 in Fig. 4.3) can then be utilized at the H_2 valley, being its specifications compatible with this application.

Table 4.2 shows the power consumption of the process shown in Fig. 4.3. In this specific case, the heat exchange is achieved by coupling the streams involved in the process, eliminating the need for external heating/cooling utilities.

For the allocation of H_2 to HRS and considering the conversion of ammonia to hydrogen at the arrival terminal, all the ammonia stored at the arrival port is converted to hydrogen. The process scheme coincides with the one described above, with the addition of H_2 compression up to 320 bar, in line with existing technologies for transporting compressed hydrogen via truck, together with further compression up to 900 bar at each refuelling station. The energy balance of the process, depicted in Fig. 4.4, is reported in Table 4.3.

4.3.2 Decentralized Cracking

Regarding the H_2 destination to HRS and considering the conversion of ammonia to hydrogen at the end users, the ammonia stored at the arrival port is transported to all end users, where its reconversion to hydrogen takes place.

The decentralized cracking process, as simulated in Aspen Plus V11®, is shown in Fig. 4.5. The process is similar to the one reported in Sect. 3.1, except that the cracking reaction is powered by electricity, in line with commercially available small-scale equipment for the decomposition of ammonia to hydrogen. Therefore, the combustion section for heat generation is not necessary. Downstream of the reaction stage, the separation of the produced hydrogen from the unconverted ammonia and the nitrogen is accomplished via PSA. The separated nitrogen stream (stream WASTE GAS in Fig. 4.5), containing hydrogen at approximately 30% by moles, needs to be appropriately disposed of (for example, by flaring combustion). The outlet hydrogen, GH_2 in Fig. 4.5, can then be utilized, after compression to 900 bar, by the end users, as its specifications are compatible with this application.

The power consumptions of the process, including the contributions for the compression of hydrogen above the HRS operating pressure, are reported in Table 4.4.

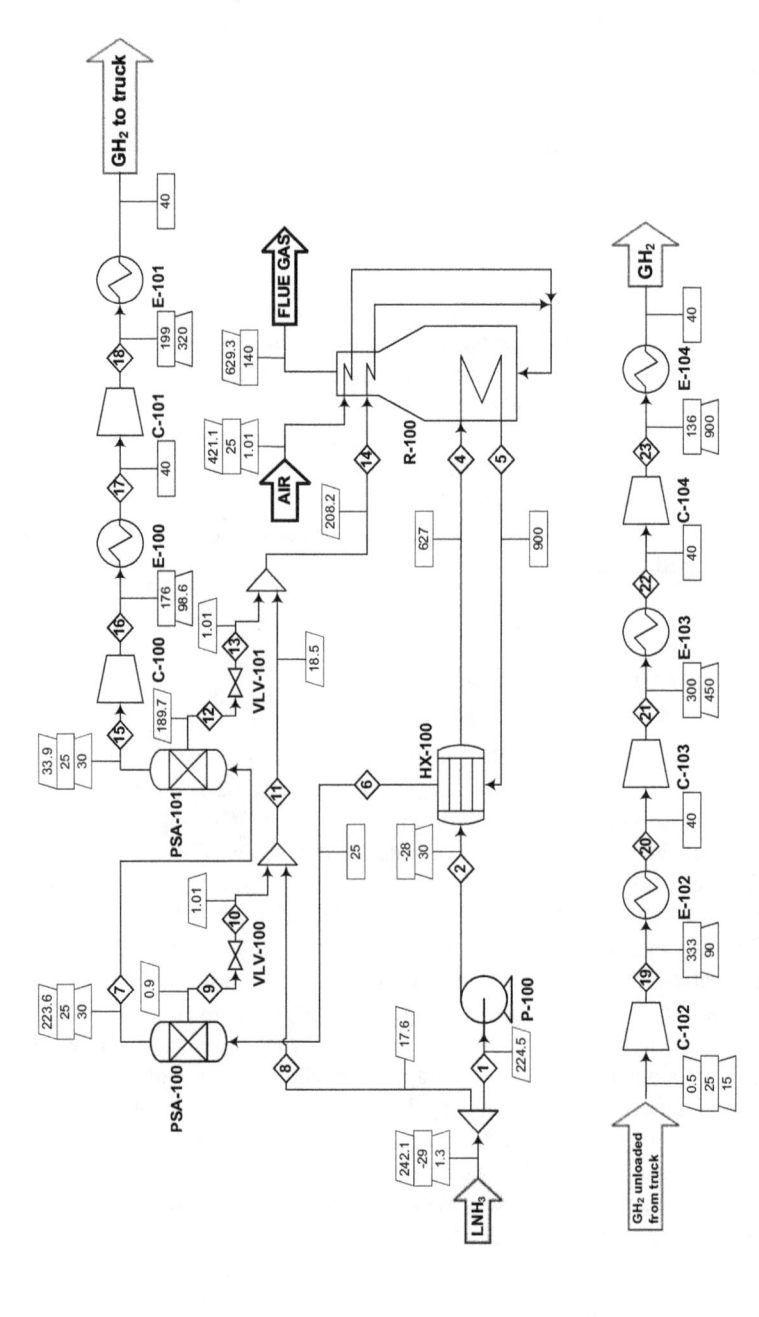

Fig. 4.4 PFD of the centralized ammonia cracking process at the arrival port and decentralized hydrogen compression process at the HRS

Table 4.3 Energy balance of the process in Fig. 4.4

Thermal energy		Electric energy	
Equipment	Q [kW]	Equipment	W [kW]
E-100	−778.5	P-100	23.9
E-101	−922.4	C-100	861.6
E-102	−1749.7 ($N_{HRS} \bullet Q_{E\text{-}102}$)	C-101	947.2
E-103	−1812.2 ($N_{HRS} \bullet Q_{E\text{-}103}$)	C-102	1887.8 ($N_{HRS} \bullet W_{C\text{-}102}$)
E-104	−736.0 ($N_{HRS} \bullet Q_{E\text{-}104}$)	C-103	1910.3 ($N_{HRS} \bullet W_{C\text{-}103}$)
		C-104	879.8 ($N_{HRS} \bullet W_{C\text{-}104}$)

4.4 Techno-Economic Assessment

Figure 4.6 shows the BFD of the ammonia value chain for the H_2 valley case and for the HRS case. In both cases, losses of carrier occur due to boil-off during sea transport. In case of hydrogen delivery to a H_2 valley and to refuelling stations with centralized cracking, the carrier is utilized as fuel to provide the heat of reaction in the reconversion process. On the contrary, in case of hydrogen delivery to refuelling stations with decentralized cracking, the heat of reaction for the cracking process is supplied by electricity. Therefore, a higher number of end users can be served in this case, about 74 in the decentralized configuration with respect to 68 in the centralized one.

The economic evaluations carried out for the ammonia value chain, according to the methodology described in Chap. 2, are reported below. The following assumptions are made:

- For the ammonia synthesis section, the fixed costs related to the refrigerant used to liquefy the ammonia are neglected;
- For the cracking section, the fixed costs of the PSA aimed at the NH_3 separation from N_2 and H_2 are neglected.

Furthermore, in both the centralized and decentralized conversions, the costs of the cracking unit are from the literature, to allow for an estimate as realistic as possible. In particular, for centralized cracking the cost function reported by Cesaro et al. [4] is used, while for cracking at the end user the estimate suggested by Thermal Dynamix™ [5], which markets such equipment, is introduced.

The fixed costs and the operating costs for the ammonia synthesis and cracking processes are evaluated based on the process simulations detailed in Sects. 4.2 and 4.3.

As regards the synthesis of ammonia, the economic evaluations provided a value of fixed costs equal to 91.12 M€ (this value includes the nitrogen separation section).

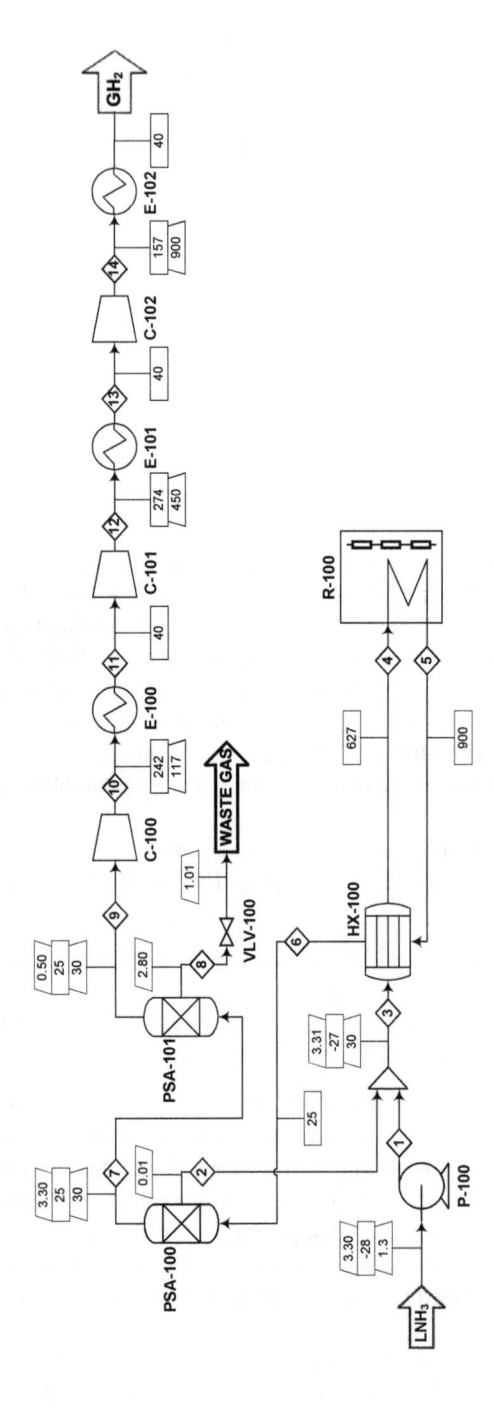

Fig. 4.5 PFD of the decentralized ammonia cracking process at the HRS

4.4 Techno-Economic Assessment

Table 4.4 Energy balance of the process in Fig. 4.5

Thermal energy		Electric energy	
Equipment	Q [kW]	Equipment	W [kW]
E-100	−1249.4 ($N_{HRS} \cdot Q_{E\text{-}100}$)	P-100	15.1 ($N_{HRS} \cdot W_{P\text{-}100}$)
E-101	−1474.2 ($N_{HRS} \cdot Q_{E\text{-}101}$)	C-100	1362.1 ($N_{HRS} \cdot W_{C\text{-}100}$)
E-102	−749.3 ($N_{HRS} \cdot Q_{E\text{-}102}$)	C-101	1567.4 ($N_{HRS} \cdot W_{C\text{-}101}$)
		C-102	897.7 ($N_{HRS} \cdot W_{C\text{-}102}$)
		R-100	11,792.4 ($N_{HRS} \cdot W_{R\text{-}100}$)

The breakdown of the bare module cost is shown in Fig. 4.7, from which it is clear that the fixed costs for the synthesis process are due, mostly, to the reaction unit, exchangers and compressors.

For the synthesis stage, Fig. 4.8 shows a comparison with the fixed costs available in the literature. As can be seen, the costs of the process simulated in Aspen Plus V11® respond to the scaling law obtained from the data of the Danish Energy Agency [6] (green curve in Fig. 4.8).

The operating costs are equal to 37.37 M€/y and are mostly due to the cost of the electricity needed to power the compressors.

As regards the centralized cracking of ammonia, the fixed costs are equal to 44.15 M€. The C_{BM} cost items are shown in Fig. 4.9, from which it is clear that almost all the fixed costs are due to the cracker.

The operating costs of the centralized ammonia cracking process are equal to 10.36 M€/y. Among the utilities necessary for the process, only the electricity required by the pump to feed the ammonia supplied to the process (24 kW) is considered, the cost of which is negligible. In this case, raw materials are not necessary, as the reagent ammonia is produced in the synthesis stage. In the carried-out evaluations, the cost associated with NO_x emissions downstream of the combustion reaction is neglected, assuming they could be reduced with a non-catalytic process based on ammonia.

When allocating the hydrogen to the HRS and reconverting the ammonia into hydrogen at the port, the *CAPEX* and *OPEX* reported above must be increased by the fixed and operating costs due to the compression of the hydrogen produced for its transport in gaseous phase (about 320 bar). These compression costs are equal to, respectively, 9.72 M€ and 11.99 M€/y. Once it reaches its destination, the hydrogen must be further compressed for its use by the end user, with costs similar to those reported for the liquefied hydrogen value chain.

In the case of decentralized cracking at the end user, the fixed costs of a single unit dedicated to the production of 500 kg/d of H_2 are equal to 0.54 M€. The operating

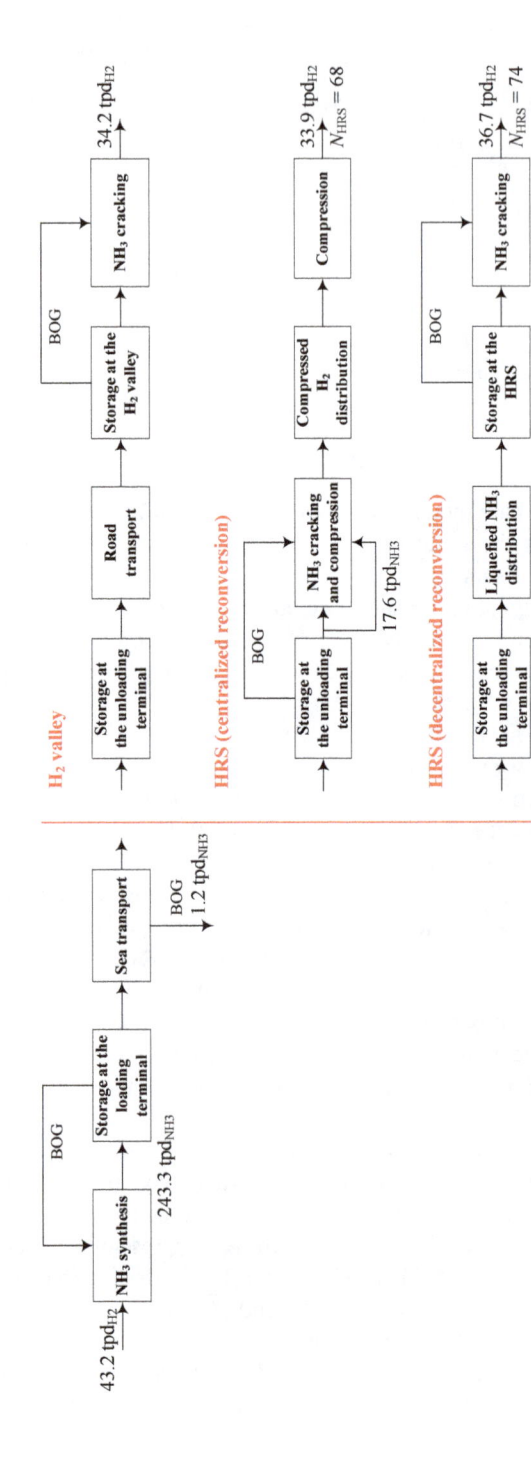

Fig. 4.6 BFD of the NH$_3$ value chain for the base case ($D = 2500$ km)

4.4 Techno-Economic Assessment

Fig. 4.7 Breakdown of the bare module cost of equipment for the ammonia synthesis process

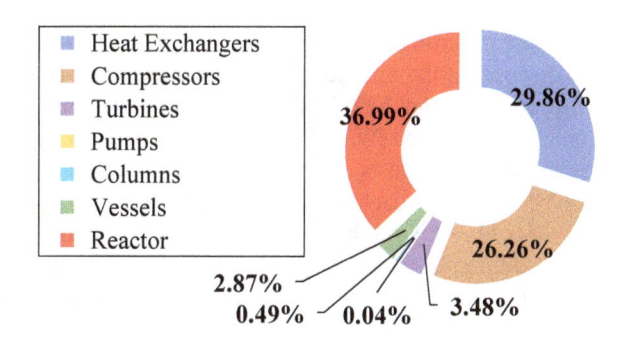

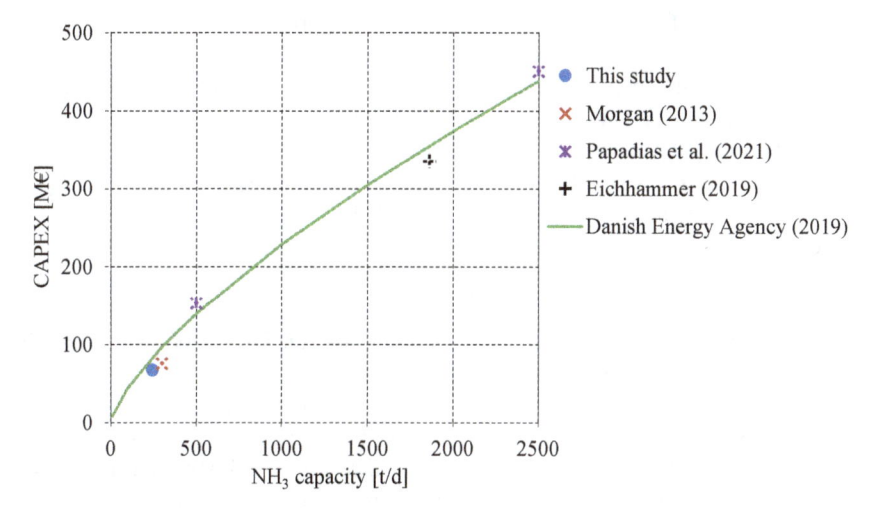

- Heat Exchangers: 29.86%
- Compressors: 26.26%
- Turbines: 3.48%
- Pumps: 0.04%
- Columns: 0.49%
- Vessels: 2.87%
- Reactor: 36.99%

Fig. 4.8 CAPEX [M€] of the ammonia synthesis section: comparison between the results obtained in this study and the literature [6–9]

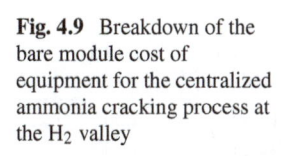

- This study
- Morgan (2013)
- Papadias et al. (2021)
- Eichhammer (2019)
- Danish Energy Agency (2019)

Fig. 4.9 Breakdown of the bare module cost of equipment for the centralized ammonia cracking process at the H_2 valley

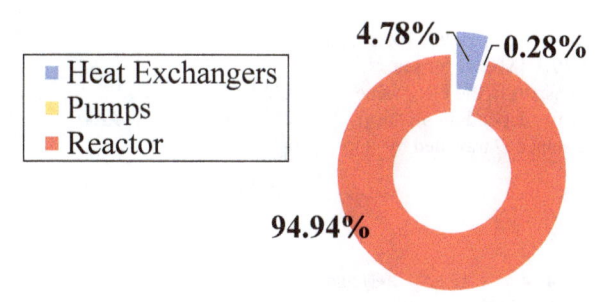

- Heat Exchangers: 4.78%
- Pumps: 0.28%
- Reactor: 94.94%

costs of the decentralized ammonia cracking process are equal to 1.24 M€/y for each end user. Among the utilities necessary for the process, only electricity is considered to power the small-scale cracker as well as the pump and compressors. Also in this case, raw materials are not necessary. In the carried-out evaluations, the cost due to the disposal of the nitrogen stream separated from the hydrogen produced through PSA, which cannot be released directly into the atmosphere due to the residual H_2 content, is neglected (WASTE GAS in Fig. 4.5). In the case of HRS with decentralized reconversion, the costs of compressing the hydrogen produced up to 900 bar must also be added. For each user, these entail a fixed cost of 1.19 M€ and an operational cost of 0.79 M€/y (in the case of compression up to 900 bar).

In addition to the cost drivers, to complete the technical–economic analysis of the value chain, the costs due to transport by ship, storage of the ammonia at the departure and arrival terminal and distribution of the produced hydrogen to the end user are estimated, depending on the alternative considered (*i.e.*, H_2 valley/HRS).

The transport of ammonia is similar to the transport of liquefied petroleum gas (LPG), as the temperature and pressure conditions for the liquid–vapour phase transition are similar: the transport of LPG is, therefore, taken as a reference. The fixed costs of transporting liquefied ammonia are obtained starting from existing case studies for LPG, considering the capacity of interest and appropriately updating them to the year 2022 via the CEPCI cost index. The operating costs of transport by ship are linked to the cost of labour (depending on the size of the crew for the ship considered), fuel consumption, CO_2 emissions, maintenance and insurance. Table 4.5 presents the fixed and operating costs for the transport of liquefied ammonia, as a function of the distance D travelled by ship.

During maritime transport, a quantity of ammonia equal to 0.1%/d is lost into the atmosphere due to the boil-off phenomenon.

The storage of liquefied ammonia is usually carried out at around -30 °C and at a pressure slightly higher than ambient pressure, 1.3 bar, inside spherical tanks equipped with polyurethane foam insulation to limit losses due to boil-off phenomenon. The fixed costs of the storage tanks, found in the literature [10–12] and appropriately adapted to the capacity of interest via Hill's law and to the year 2022 via the CEPCI index, are shown in Table 4.6.

Table 4.5 *CAPEX* [M€] and *OPEX* [M€/y] for sea transport in the NH_3 value chain, depending on the distance D travelled by ship

D [km]	V_{vessel} [m^3]	CAPEX [M€]	OPEX [M€/y]
2500 (base)	3900	19.69	4.00
5000	6500	25.37	5.49
10000	11900	34.25	7.41

Table 4.6 *CAPEX* [M€] and *OPEX* [M€/y] for one liquefied ammonia full-capacity storage tank in the NH_3 value chain, depending on the distance D travelled by ship

D [km]	V_{tank} [m^3]	CAPEX [M€]	OPEX [M€/y]
2500 (base)	4200	5.04	0.50
5000	7100	7.36	0.74
10000	13000	11.40	1.14

Table 4.7 CAPEX [M€] and OPEX [M€/y] for liquefied ammonia distribution in the NH$_3$ value chain, depending on the distance D travelled by ship

D [km]	n_{trucks}	CAPEX [M€]	OPEX [M€/y]
2500 (base)	9	3.73	2.62
5000	9	3.73	2.62
10000	9	3.73	2.62

Table 4.8 CAPEX [M€] and OPEX [M€/y] for compressed hydrogen distribution in the NH$_3$ value chain, depending on the distance D travelled by ship

D [km]	n_{trucks}	CAPEX [M€]	OPEX [M€/y]
2500 (base)	34	26.98	10.07
5000	34	26.98	10.07
10000	34	26.98	10.07

In the H$_2$ valley case and in the HRS case with cracking at the end user, liquefied ammonia is distributed on a trailer truck with a capacity of 14730 kg (22 m^3), the fixed and operating costs of which are reported in Table 4.7.

If in the HRS case ammonia cracking is carried out at the destination port, compressed hydrogen gas is distributed. In this case, the distribution costs are shown in Table 4.8 and are obtained with assumptions similar to those introduced for the LH$_2$ value chain.

References

1. Agrawal R, Thorogood R M. Production of medium pressure nitrogen by cryogenic air separation. Gas Separation & Purification. 1991; 5 December, 203–209.
2. Nielsen A, Kjaer J, Hansen B. Rate equation and mechanism of ammonia synthesis at industrial conditions. J Catal. 1964;3(1):68–79.
3. Spatolisano E, Pellegrini LA. Haber-Bosch process intensification: A first step towards small-scale distributed ammonia production. Chem Eng Res Des. 2023;195:651–61.
4. Cesaro Z. Ives M C, Nayak-Luke R, Mason M, Bañares-Alcántara R. Ammonia to power: Forecasting the levelized cost of electricity from green ammonia in large-scale power plants. Applied Energy. 2021; 282(61), 116009.
5. https://www.thermaldynamix.com/.
6. Danish Energy Agency and Energinet, 2017. Technology Data—Renewable fuels.
7. Morgan E R. Techno-Economic feasibility study of ammonia plants powered by offshore Wind. Ph.D Dissertation. University of Massachusetts Amherst. 2013.
8. Papadias DD, Peng J-K, Ahluwalia RK. Hydrogen carriers: Production, transmission, decomposition, and storage. Int J Hydrogen Energy. 2021;46(47):24169–89.
9. Eichhammer W, Oberle S, Händel M, Boie I, Gnann T, Wietschel M, Lux B. Study on the opportunities of "power-to-X" in Morocco. 10 hypotheses for discussion. 2019. Fraunhofer Institute for Systems and Innovation Research ISI, Breslauer Straße 48, 76139 Karlsruhe, Germany.

10. https://www.igu.org/wp-content/uploads/2020/12/P1-40_William-Leighty.pdf.
11. https://www.pc.gov.au/inquiries/completed/climate-change-adaptation/submissions/sub046-attachment3.pdf.
12. https://arpa-e.energy.gov/sites/default/files/03%20ARPA-E%20REFUEL%20Kickoff%20Meeting%20%20%208-16-17.pdf.

Chapter 5
Toluene/methylcyclohexane as Green H$_2$ Carrier

Abstract Liquid organic hydrogen carriers (LOHCs) are organic molecules which can undergo cyclic hydrogenation and dehydrogenation. Among all the possible choices of organic carriers, toluene is presently the most mature and extensively studied compound. The couple toluene (TOL)/methylcyclohexane (MCH) for H$_2$ transport has been patented by Chiyoda Corporation. Their "AHEAD" project marked the first demonstration of intercontinental hydrogen transport through LOHC, moving 210 tonnes of hydrogen from Brunei to Japan. The cost drivers of the toluene/methylcyclohexane value chain are the hydrogenation and dehydrogenation processes. Therefore, their detailed technical assessment is of paramount importance for establishing their reliability as H$_2$ transport alternatives. This chapter details the toluene hydrogenation and methylcyclohexane dehydrogenation sections and, considering the process simulation results, provide the economic assessment of each stage of the value chain.

Keywords LOHC · Process design · Detailed Techno-economic assessment · Hydrogenation · Dehydrogenation

5.1 Introduction

Liquid Organic Hydrogen Carriers (LOHCs) are organic molecules that can be reversibly hydrogenated and dehydrogenated to release hydrogen. Being liquid at ambient temperature and pressure, they can effectively store hydrogen without losses. Additionally, their oil-like properties allow them to be easily handled using existing oil infrastructure. With growing research interest in these compounds, numerous reviews aim to identify the best candidates for hydrogen storage and transport [1–3]. The ideal LOHC should exhibit: a low melting point to prevent solidification issues; a low dehydrogenation enthalpy to minimize the energy required for dehydrogenation; a high boiling point to prevent volatilization and facilitate separation from the produced hydrogen; a high hydrogen storage capacity; low toxicity and low cost.

Fig. 5.1 Toluene value chain

Several cyclic hydrocarbons could serve as LOHCs, including benzene/cyclohexane, toluene/methylcyclohexane, naphthalene/decalin, biphenyl/bicyclohexyl, and dibenzyltoluene/perhydro-dibenzyltoluene.

Among them, toluene (TOL)/methylcyclohexane (MCH) is the most mature technology, firstly proposed by the Chiyoda corporation [4] and already available at the demonstration scale.

For this reason, it is selected as the benchmark, to understand the performances of all the other organic couples when compared to this alternative.

The toluene value chain shown in Fig. 5.1, includes the following steps:

- *Hydrogenation.* Green hydrogen coming from the battery limits is chemically bonded in toluene (TOL), to produce methylcyclohexane (MCH).
- *Storage at the departure terminal.* Methylcyclohexane is stored in atmospheric tanks, similar to the ones used for oil products.
- *Maritime transport.* Seaborne transport occurs via ships similar to the oil product ones.
- *Storage at the arrival terminal.* Similarly to the storage at the departure terminal, storage takes place into atmospheric tanks.
- *Dehydrogenation.* Methylcyclohexane is dehydrogenated to toluene, to favour H_2 release. Toluene can be, thus, recycled back to the loading terminal.
- *Distribution.* Distribution depends on the H_2 final destination, which can be either H_2 valley or hydrogen refuelling stations. In the first case, the hydrogenated organic carrier is routed via truck to the H_2 valley, while for mobility sector applications, gaseous hydrogen is distributed, having been compressed to increase its density and stored in appropriate cylindrical tanks (tube trailers), via trailer trucks.

5.2 Toluene Hydrogenation

Figure 5.2 illustrates the process simulation for toluene hydrogenation. Toluene from the unloading terminal is combined with a make-up stream, then pumped and heated to the reaction conditions: 20 bar and 210 °C. Heating occurs in a process-process heat exchanger HX-100 to exploit the high enthalpy content of the reaction products. The hydrogenation reaction 5.1 which takes place in R-100 is highly exothermic, so nitrogen is introduced as a thermal diluent to control the reaction's exothermic nature. The nitrogen flow rate is adjusted to maintain a concentration of about 17 mol% in

5.3 Methylcyclohexane Dehydrogenation

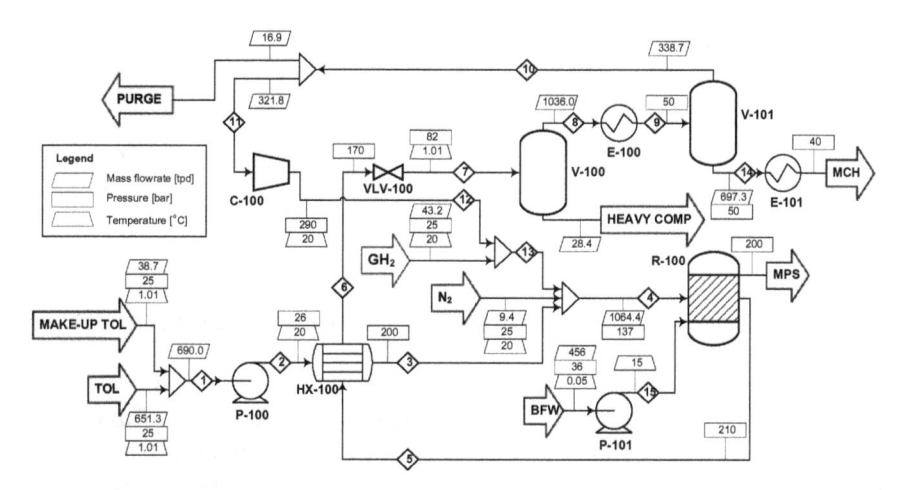

Fig. 5.2 PFD of the toluene hydrogenation process

the reactor R-100 inlet stream [5]. Alongside nitrogen and toluene, green hydrogen produced from water electrolysis (stream GH$_2$ in Fig. 5.2) is supplied to the system. The reactor is simulated using the RGibbs module of Aspen Plus®, where reactants conversion follows thermodynamic equilibrium. The mixture leaving the reactor (stream 5 in Fig. 5.2), containing unconverted hydrogen, nitrogen, and methylcyclohexane, is cooled in HX-100 and then directed to downstream separation stages for product and by-product recovery. The separation process includes two flash drums in series (V-100 and V-101), where the reactor outlet mixture, after passing through VLV-100 valve for expansion, is purified first from heavy ends and then from light ends. The produced methylcyclohexane (stream MCH in Fig. 5.2) is stored and shipped from the unloading terminal, while unconverted hydrogen (stream 10 in Fig. 5.2) is compressed and recycled back to the reaction section. The reaction heat is exploited for steam production from boiler feed water, as pointed out in Fig. 5.2. Inlet boiler feed water (stream BFW in Fig. 5.2) is pumped in P-101 and then routed to a heat exchanger for its vaporization.

$$C_7H_8 + 3H_2 \rightarrow C_7H_{14} \tag{5.1}$$

The energy balance for the process depicted in Fig. 5.2 is reported in Table 5.1.

5.3 Methylcyclohexane Dehydrogenation

The dehydrogenation process for toluene regeneration from methylcyclohexane is depicted in Fig. 5.3. Initially, methylcyclohexane is pumped and directed to the reaction section. The reactor R-100 operates at 350 °C and 3 bar. Since its reaction

Table 5.1 Energy balance of the process in Fig. 5.2

Thermal energy		Electric energy	
Equipment	Q [kW]	Equipment	W [kW]
E-100	−3378.1	P-100	30.2
E-101	−152.4	P-101	15.3
		C-100	1412.5

rate expression is not clearly assessed in literature [5], it is modelled as a black box with fixed conversion.

Downstream the reactor, the mixture exiting R-100 contains toluene, hydrogen, and unwanted by-products (mainly benzene, formed from reaction 5.3). This mixture is then sent to the downstream separation train. The toluene-rich stream (stream TOL in Fig. 5.3) needs to be stored and redirected to the loading terminal. Meanwhile, hydrogen undergoes purification via pressure swing adsorption PSA-100 to meet the necessary purity standards.

$$C_7H_{14} \rightarrow C_7H_8 + 3H_2 \tag{5.2}$$

$$C_7H_8 + H_2 \rightarrow C_6H_6 + CH_4 \tag{5.3}$$

Because the dehydrogenation reaction is highly endothermic, its required heat is provided by burning a portion of the generated hydrogen along with waste streams leaving the system (streams 26 and 27 in Fig. 5.3). A portion of the hydrogen, along with waste streams and the necessary air for combustion, undergoes preheating through heat exchange with flue gas until it reaches its auto-ignition temperature (around 500 °C). The hydrogen flow rate is adjusted to balance the heat generated by combustion with that needed for dehydrogenation, while the air flow rate is slightly higher than stoichiometric to ensure complete oxidation of inlet nitrogen. Consequently, the significant endothermic nature of the dehydrogenation reaction results in a reduced amount of hydrogen leaving the system (Table 5.2).

5.4 Techno-Economic Assessment

Figure 5.4 shows the BFD of the toluene value chain for the H_2 valley case and for the HRS case.

The mass balance for each block of the value chain is outlined. As can be observed, about 1/3 of the H_2 coming from the electrolysers is lost in the dehydrogenation section, to provide the heat necessary for the reaction to occur. This is particularly evident for the H_2 destination to HRS: in this case, only 53 end users can be served.

5.4 Techno-Economic Assessment

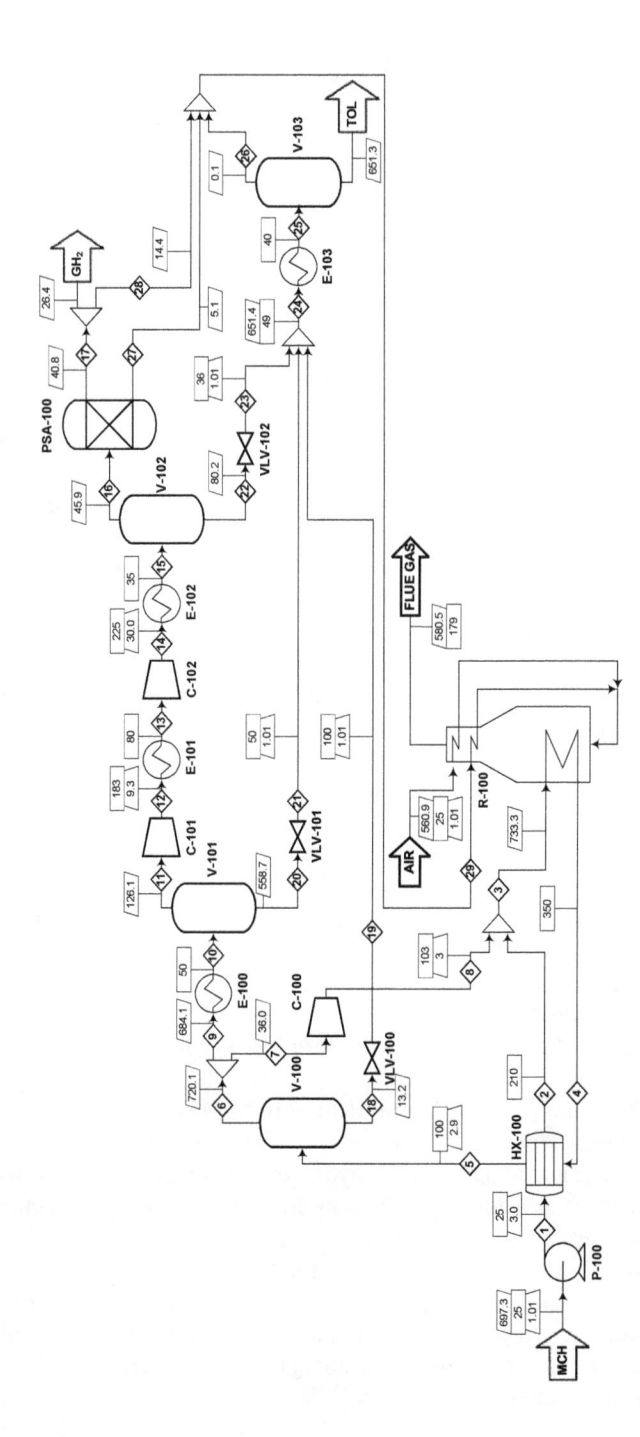

Fig. 5.3 PFD of the methylcyclohexane dehydrogenation process

Table 5.2 Energy balance of the process in Fig. 5.3

Thermal energy		Electric energy	
Equipment	Q [kW]	Equipment	W [kW]
E-100	−3344.6	P-100	3.5
E-101	−872.4	C-100	2.42
E-102	−1951.9	C-101	1113.3
E-103	−119.8	C-102	1231.2

The techno-economic assessment of toluene value chain has been carried out according to the methodology described in Chap. 2.

For both cases, the fixed costs associated with the hydrogenation/dehydrogenation reactor have been assessed based on those of a shell and tube exchanger with equivalent duty. This approach stems from the current limitations in accurately sizing the equipment during this stage of analysis. Additionally, the economic evaluations undertaken also include the fixed cost of the initial loading of LOHC. This initial loading volume is assumed to be equivalent to that of three storage tanks, representing the amount of carrier required for the start-up of H_2 delivery, which is variable at variable distance D travelled by ship, as pointed out in Table 5.3.

Toluene hydrogenation *CAPEX* are estimated to be 11.17 M€. More than half of the fixed costs are due to the non-converted hydrogen recycle compressor (see Fig. 5.5a). This unit is necessary for the recompression to the reaction pressure of the hydrogen expanded to ambient pressure in the separation section.

The operating costs of toluene hydrogenation are 8.57 M€/y. Among the utilities, electricity is the main cost item. The steam produced from boiler feed water, exploiting the reaction duty, has been considered as a revenue. Among the raw materials necessary for the toluene hydrogenation section, the make-up toluene and the nitrogen entering the reactor as a thermal diluent are considered. Most of the expenditure for raw materials is due to the toluene make up, necessary to compensate for the losses in the separation sections downstream of the reactor.

As regards the dehydrogenation process, most of the fixed costs, equal to 17.21 M€, are due to the compressors necessary for H_2 compression up to 30 bar, as from Fig. 5.5b.

Concerning the operating costs, they result to be 18.88 M€/y. The primary expense is electricity for powering pumps and compressors. There is no need for raw materials, as methylcyclohexane is supplied from the hydrogenation process. In the evaluations performed, the cost related to CO_2 emissions from the combustion reaction was disregarded, as it turns out to be negligible.

In addition to the cost drivers, to complete the techno-economic analysis of the value chain, the costs associated with shipping the carrier, storing the hydrogenated and dehydrogenated carrier at the departure and arrival terminals, and distributing the produced hydrogen to the end user, depending on the alternative considered (*i.e.*, H_2 valley/HRS), are estimated.

5.4 Techno-Economic Assessment

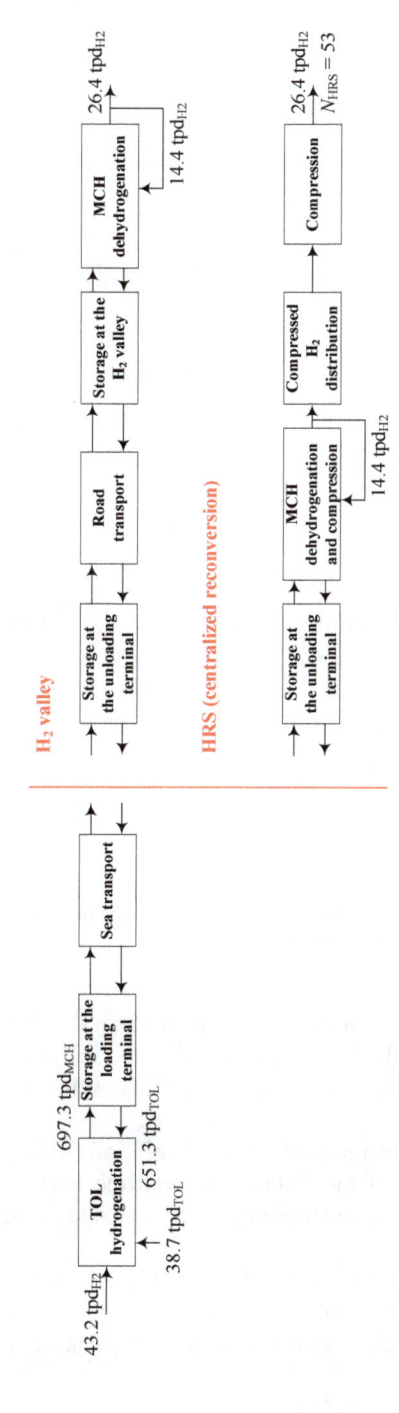

Fig. 5.4 BFD of the toluene value chain for the base case ($D = 2500$ km)

Table 5.3 CAPEX [M€] of the toluene initial loading depending on the distance D travelled by ship

D [km]	CAPEX [M€]
2500	27.94
5000	47.50
10000	86.62

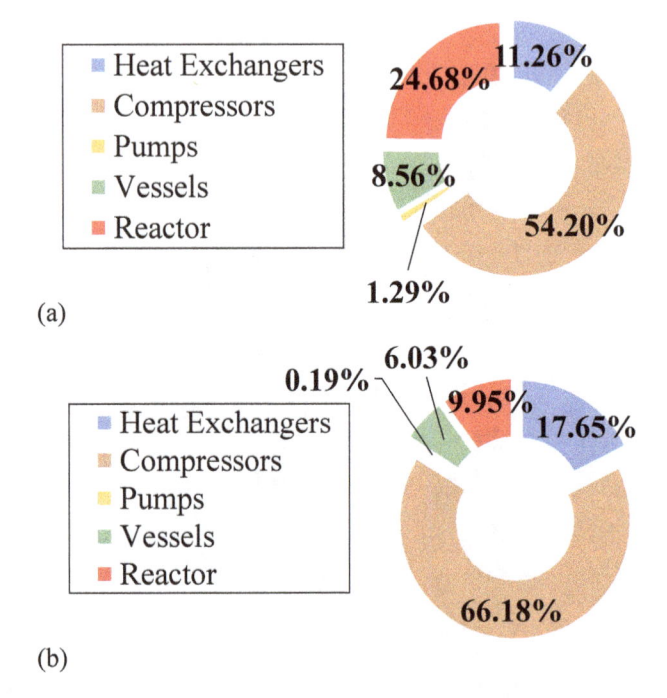

Fig. 5.5 Breakdown of the bare module cost of equipment for: (**a**) toluene hydrogenation and (**b**) methylcyclohexane dehydrogenation process

The costs of storage tanks at the departure and arrival terminals are evaluated based on data reported by the IEA [6], adjusted to the relevant capacity using the Hill's equation and updated to 2022 using the CEPCI cost index, and are detailed in Table 5.4.

Table 5.5 summarizes ship transport *CAPEX* [M€] and *OPEX* [M€/y] for toluene value chain, as a function of the distance D from the loading to the unloading terminal. These costs are evaluated referring to the available literature for oil product applications [7].

The results for the transport section in the H_2 valley case are reported in Table 5.6 and refer to the LOHC distribution on a trailer truck with a capacity of 28500 kg (37 m^3), whose fixed costs are taken from Teichman et al. [8] and adjusted for inflation.

Table 5.4 CAPEX [M€] and OPEX [M€/y] for one TOL/MCH full-capacity storage tank in the toluene value chain, depending on the distance D travelled by ship

D [km]	Stored compound	V_{tank} [m³]	CAPEX [M€]	OPEX [M€/y]
2500 (base)	MCH	10,300	14.04	1.40
	TOL	8500	12.51	1.25
5000	MCH	17400	19.23	1.92
	TOL	14400	17.17	1.72
10,000	MCH	31700	27.56	2.76
	TOL	26200	24.59	2.46

Table 5.5 CAPEX [M€] and OPEX [M€/y] for sea transport in the toluene value chain, depending on the distance D travelled by ship

D [km]	V_{vessel} [m³]	CAPEX [M€]	OPEX [M€/y]
2500 (base)	9300	10.20	4.79
5000	15800	13.59	6.91
10000	28800	19.00	10.15

Table 5.6 CAPEX [M€] and OPEX [M€/y] for TOL/MCH distribution in the toluene value chain, depending on the distance D travelled by ship

D [km]	n_{trucks}	CAPEX [M€]	OPEX [M€/y]
2500 (base)	13	5.23	4.03
5000	13	5.23	4.03
10000	13	5.23	4.03

Table 5.7 CAPEX [M€] and OPEX [M€/y] for compressed hydrogen distribution in the toluene value chain, depending on the distance D travelled by ship

D [km]	n_{trucks}	CAPEX [M€]	OPEX [M€/y]
2500 (base)	27	25.19	9.86
5000	27	25.19	9.86
10000	27	25.19	9.86

For the HRS, dehydrogenation is performed at the arriving terminal. Thus, compressed gaseous hydrogen is to be distributed, whose associated costs are outlined in Table 5.7.

References

1. Reuß M, Grube T, Robinius M, Preuster P, Wasserscheid P, Stolten D. Seasonal storage and alternative carriers: A flexible hydrogen supply chain model. Appl Energy. 2017;200:290–302.

2. Papadias DD, Peng J-K, Ahluwalia RK. Hydrogen carriers: Production, transmission, decomposition, and storage. Int J Hydrogen Energy. 2021;46:24169–89.
3. Noh H, Kang K, Seo Y. Environmental and energy efficiency assessments of offshore hydrogen supply chains utilizing compressed gaseous hydrogen, liquefied hydrogen, liquid organic hydrogen carriers and ammonia. Int J Hydrogen Energy. 2023;48:7515–32.
4. https://www.chiyodacorp.com/en/service/spera-hydrogen/.
5. Imagawa K, Kawai H, Shiraga M, Nakajima Y. Aromatic compound hydrogenation system and hydrogen storage/transport system equipped with same, and aromatic compound hydrogenation method. WO2015115101A1, 2015.
6. IEA G20 Hydrogen report: Assumptions. 2019.
7. https://compassmar.com.
8. Teichmann D, Arlt W, Wasserscheid P. Liquid Organic Hydrogen Carriers as an efficient vector for the transport and storage of renewable energy. Int J Hydrogen Energy. 2012;37:18118–32.

Chapter 6
Dibenzyltoluene/ Perhydro-Dibenzyltoluene as Green H$_2$ Carrier

Abstract A variety of liquid organic hydrogen carriers (LOHCs) have been proposed in literature, with the aim of enabling long-distance H$_2$ transport. Together with toluene, which is the most mature technology, dibenzyltoluene has been regarded as a promised candidate to overcome some drawbacks of the couple toluene/ methylcyclohexane, as the coproduction of benzene in the reaction section. To analyse the feasibility of H$_2$ transport through dibenzyltoluene, this chapter details the dibenzyltoluene value chain. Hydrogenation and dehydrogenation sections have been analysed in detail thanks to the Aspen Plus V11® process simulation. Considering the process simulation results, the economic assessment of each stage of the value chain is provided.

Keywords LOHC · Process design · Detailed Techno-economic assessment · Hydrogenation · Dehydrogenation

6.1 Introduction

Dibenzyltoluene (DBT-H0), that in its hydrogenated form becomes perhydrodibenzyltoluene (DBT-H18), is a homocyclic organic compound proposed by Hydrogenious for large-scale H$_2$ transport [1, 2]. Despite regarded as promising from the available literature, dibenzyltoluene is approximately ten times more expensive than toluene, significantly impacting the overall process economics. Further research is needed to enable this alternative cost-effective. For this reason, in the following sections, a detailed techno-economic assessment of the H$_2$ transport through dibenzyltoluene is performed, to pave the way for future process intensification.

The dibenzyltoluene value chain shown in Fig. 6.1 includes the following steps:

- *Hydrogenation.* Green hydrogen coming from the battery limits is chemically bonded in dibenzyltoluene, to produce perhydro-dibenzyltoluene.
- *Storage at the departure terminal.* Perhydro-dibenzyltoluene is stored in atmospheric tanks, similar to the ones used for oil products.

Fig. 6.1 Dibenzyltoluene value chain

- *Maritime transport.* Seaborne transport occurs via ships similar to the oil product ones.
- *Storage at the arrival terminal.* Similarly to the storage at the departure terminal, storage takes place into atmospheric tanks.
- *Dehydrogenation.* Perhydro-dibenzyltoluene is dehydrogenated to dibenzyltoluene, to favour H_2 release. Dibenzyltoluene can be, thus, recycled back to the loading terminal.
- *Distribution.* Distribution depends on the H_2 final destination, which can be either H_2 valley or hydrogen refuelling stations. In the first case, the hydrogenated organic carrier is routed via truck to the H_2 valley, while for mobility sector applications, gaseous hydrogen is distributed, having been compressed to increase its density and stored in appropriate cylindrical tanks (tube trailers), via trailer trucks.

Before the process simulation phase, dibenzyltoluene and its corresponding hydrogenated form, not available in the Aspen Plus databank, have been added as pseudo-components, considering the physical properties available in literature, together with the expressions for the heat capacity, vapour pressure and density evaluation (see Spatolisano et al. [3] for further details).

6.2 Dibenzyltoluene Hydrogenation

The dibenzyltoluene hydrogenation process is represented in Fig. 6.2. The dehydrogenated organic carrier stored at the loading terminal, (stream DBT-H0 in Fig. 6.2) and the make-up stream (whose flow rate is very limited) are routed to the reactor, after being pumped and heated. The reactor R-100 is fed with dibenzyltoluene, hydrogen and nitrogen. The hydrogenation reaction is:

$$C_{21}H_{20} + 9H_2 \rightarrow C_{21}H_{38} \tag{6.1}$$

The reactor operates at 35 bar and 210 °C. Due to the current state of DBT hydrogenation and dehydrogenation technology, a detailed kinetic scheme cannot be applied directly to process simulations. Instead, experimental conversion and selectivity data are integrated into the simulation phase to address this challenge. For instance, Shi et al. [4] showed complete DBT-H0 hydrogenation using a 5 wt.% Pt/Al_2O_3 catalyst at around 200 °C and pressures between 30 and 40 bar. This

6.2 Dibenzyltoluene Hydrogenation

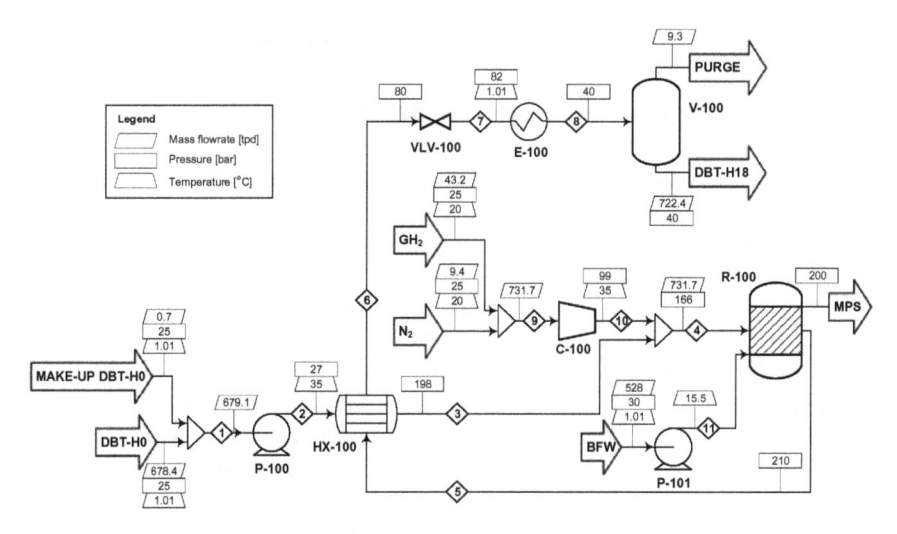

Fig. 6.2 PFD of the dibenzyltoluene hydrogenation process

information guided the assumption of complete hydrogen conversion in the reactor, simplifying the process by avoiding a recycle loop.

The resulting hydrogenated product is cooled, stored, and then transported to the unloading terminal. Downstream of the reactor, the separation process is simplified due to the quantitative hydrogen conversion and minimal by-product formation. This simplicity is recognized in literature as a significant advantage of dibenzyltoluene as a hydrogen carrier. Additionally, because dibenzyltoluene is heavier than toluene, its separation benefits from better vapour-liquid equilibrium exploitation, owing to larger differences in relative volatilities with lighter compounds. Furthermore, the heat generated from the reaction is utilized for steam production from boiler feed water, contributing to process efficiency and sustainability.

Table 6.1 details the energy balance for the process of Fig. 6.2.

Table 6.1 Energy balance of the process in Fig. 6.2

Thermal energy		Electric energy	
Equipment	Q [kW]	Equipment	W [kW]
E-100	-623.4	P-100	47.2
		P-101	16.8
		C-100	550.6

6.3 Perhydro-Dibenzyltoluene Dehydrogenation

As regards the DBT-H18 dehydrogenation to DBT-H0 (reaction 6.2), its process flow diagram is reported in Fig. 6.3.

The DBT-H18 entering the battery limits is pumped and routed to the reaction section, where the dehydrogenation reaction 6.2 takes place. The reactor R-100 works at 320 °C and 1.1 bar, with fixed conversion as reported in literature [5].

$$C_{21}H_{38} \rightarrow C_{21}H_{20} + 9H_2 \quad (6.2)$$

The outlet product mixture, composed of DBT and hydrogen, undergoes a sequence of separations downstream the reaction section. The outlet DBT (stream DBT-H0 in Fig. 6.3) has to be stored and routed back to the loading terminal, while the produced hydrogen, essentially pure, is compressed and routed to the hydrogen valley.

Since the dehydrogenation reaction is highly endothermic, the heat of reaction is supplied by burning part of the hydrogen produced (stream 26 in Fig. 6.3). This stream is routed to the combustion section. In this section, the hydrogen stream is preheated by heat exchange with the flue gas up to its auto-ignition temperature (about 500 °C) and then burned with an air flow rate slightly higher than the stoichiometric one, to ensure complete oxidation of the fuel. The hydrogen flow rate is modulated in such a way that the heat generated by the combustion is equal to that necessary for the dehydrogenation reaction.

The energy balance of the process of Fig. 6.3 is reported in Table 6.2, considering both thermal and electric energy requirements.

6.4 Techno-Economic Assessment

Figure 6.4 reports the block flow diagram for dibenzyltoluene value chain, where the mass balance for each step is outlined, for both the hydrogen destinations to H_2 valley or HRS.

As observed for toluene in Chap. 5, about 1/3 of the H_2 coming from the electrolyser is lost in the dehydrogenation section, to provide the heat necessary for the reaction to occur. For the H_2 destination to HRS, in this case, 56 end users can be served.

Fixed and operating costs are estimated for the processes of hydrogenation and dehydrogenation of dibenzyltoluene based on detailed process simulations in Sects. 6.2 and 6.3, respectively. For the hydrogenation process, *CAPEX* are 8.98 M€, with approximately one-third of these costs attributed to the incoming hydrogen compressor, as detailed in Fig. 6.5a. These fixed costs are comparatively lower than those for toluene hydrogenation, reflecting the simplified plant layout downstream of the reaction section.

6.4 Techno-Economic Assessment

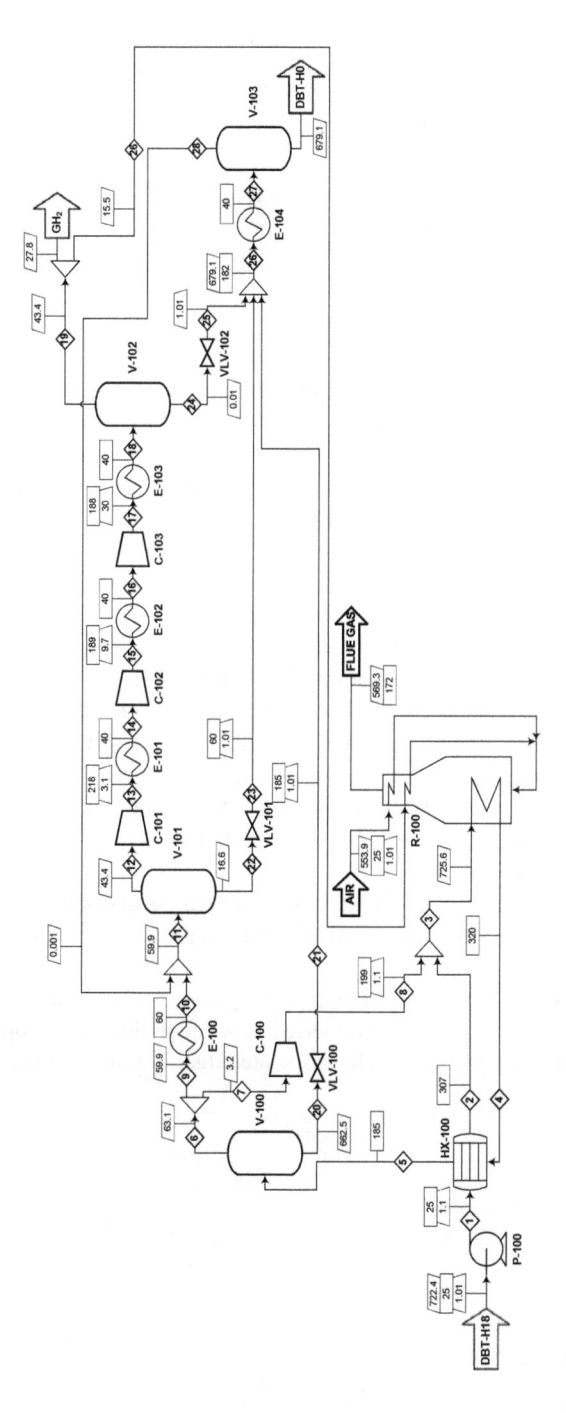

Fig. 6.3 PFD of the perhydro-dibenzyltoluene dehydrogenation process

Table 6.2 Energy balance of the process in Fig. 6.3

Thermal energy		Electric energy	
Equipment	Q [kW]	Equipment	W [kW]
E-100	−1002.3	P-100	0.14
E-101	−1291.3	C-100	5.75
E-102	−1080.4	C-101	1148.0
E-103	−1076.7	C-102	1081.5
E-104	−1974.7	C-103	1080.2

OPEX for the dibenzyltoluene hydrogenation process amount to 9.17 M€/y, inclusive of utilities such as electricity for pumps and compressors, cooling water, and boiler feed water. The generated steam is not accounted for as revenue due to the preliminary nature of the simulation. Raw materials include dibenzyltoluene make-up and nitrogen as a thermal diluent, with the make-up costs due to losses in downstream separation being notably higher than those for toluene. However, the make-up flow rate of dibenzyltoluene remains significantly lower than that of toluene in its respective value chain. Nevertheless, due to the higher costs of dibenzyltoluene as raw material, the capital expenses of initial loading needed to enable the steady-state cyclic operation are far from negligible, as detailed in Table 6.3.

On the other hand, the dehydrogenation of DBT-H18 incurs *CAPEX* of 21.40 M€ and *OPEX* of 25.71 M€/y, primarily due to the compressors needed to compress hydrogen up to 30 bar, aligning with its operating pressure in the H_2 valley, as outlined in Fig. 6.5b.

Similarly to toluene as hydrogen carrier, already detailed in Chap. 5, Tables 6.4 and 6.5 show, respectively, the costs related to sea transport and storage for the case of dibenzyltoluene as LOHC.

For H_2 industrial application (H_2 valley), the LOHC is distributed via trucks of capacity of 28500 kg (31 m^3), for which fixed and operating costs are reported in Table 6.6.

For the Hydrogen Refuelling Station, dehydrogenation occurs at the destination port, and the task is to distribute compressed gaseous hydrogen at approximately 250 320 bar and room temperature. The associated distribution costs are outlined in Table 6.7.

6.4 Techno-Economic Assessment

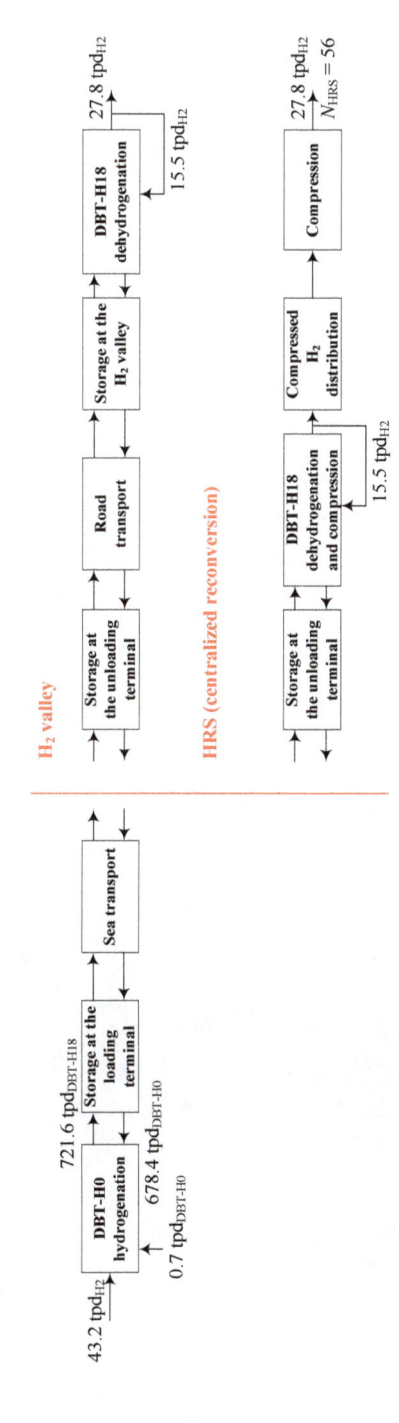

Fig. 6.4 BFD of the dibenzyltoluene value chain for the base case ($D = 2500$ km)

Fig. 6.5 Breakdown of the bare module cost of equipment for: (**a**) dibenzyltoluene hydrogenation and (**b**) perhydro-dibenzyltoluene dehydrogenation process

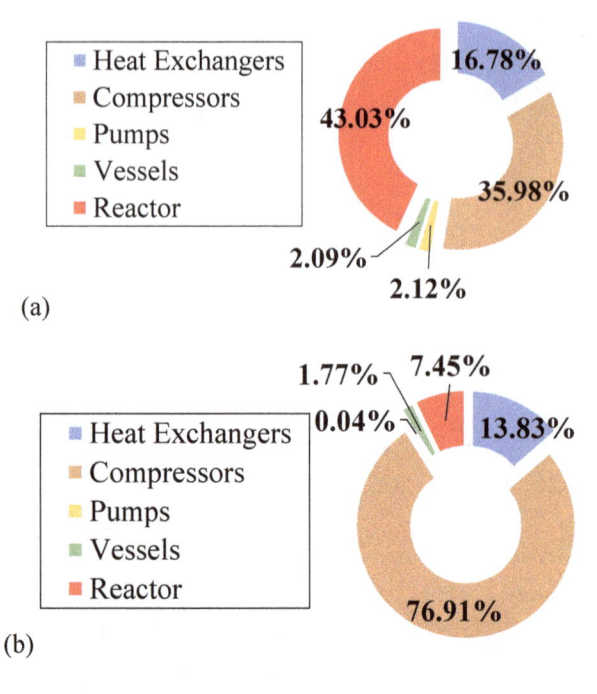

(a) Heat Exchangers 16.78%, Compressors 43.03%, Pumps 2.09%, Vessels 2.12%, Reactor 35.98%

(b) Heat Exchangers 13.83%, Compressors 76.91%, Pumps 0.04%, Vessels 1.77%, Reactor 7.45%

Table 6.3 CAPEX [M€] of the dibenzyltoluene initial loading depending on the distance D travelled by ship

D [km]	CAPEX [M€]
2500	111.94
5000	190.29
10000	347.00

Table 6.4 CAPEX [M€] and OPEX [M€/y] for sea transport in the dibenzyltoluene value chain, depending on the distance D travelled by ship

D [km]	V_{vessel} [m^3]	CAPEX [M€]	OPEX [M€/y]
2500 (base)	7600	9.17	4.23
5000	12,900	12.16	6.17
10000	23,600	16.99	9.15

Table 6.5 CAPEX [M€] and OPEX [M€/y] for one DBT-H0/H18 full-capacity storage tank in the dibenzyltoluene value chain, depending on the distance D travelled by ship

D [km]	Stored compound	V_{tank} [m^3]	CAPEX [M€]	OPEX [M€/y]
2500 (base)	DBT-H18	7900	11.98	1.20
	DBT-H0	8400	12.42	1.24
5000	DBT-H18	13400	16.44	1.64
	DBT-H0	14200	17.03	1.70
10000	DBT-H18	24400	23.56	2.36
	DBT-H0	25900	24.42	2.44

Table 6.6 CAPEX [M€] and OPEX [M€/y] for DBT-H0/H18 distribution in the dibenzyltoluene value chain, depending on the distance D travelled by ship

D [km]	n_{trucks}	CAPEX [M€]	OPEX [M€/y]
2500 (base)	13	5.23	4.08
5000	13	5.23	4.08
10000	13	5.23	4.08

Table 6.7 CAPEX [M€] and OPEX [M€/y] for compressed hydrogen distribution in the dibenzyltoluene value chain, depending on the distance D travelled by ship

D [km]	n_{trucks}	CAPEX [M€]	OPEX [M€/y]
2500 (base)	28	26.13	10.28
5000	28	26.13	10.28
10000	28	26.13	10.28

References

1. Teichmann D, Arlt W, Wasserscheid P. Liquid Organic Hydrogen Carriers as an efficient vector for the transport and storage of renewable energy. Int J Hydrogen Energy. 2012;37:18118–32.
2. Brückner N, Obesser K, Bösmann A, Teichmann D, Arlt W, Dungs J, et al. Evaluation of industrially applied heat-transfer fluids as liquid organic hydrogen carrier systems. Chemsuschem. 2014;7:229–35.
3. Spatolisano E, Restelli F, Matichecchia A, Pellegrini LA, de Angelis AR, Cattaneo S, et al. Assessing opportunities and weaknesses of green hydrogen transport via LOHC through a detailed techno-economic analysis. Int J Hydrogen Energy. 2024;52:703–17.
4. Shi L, Qi S, Qu J, Che T, Yi C, Yang B. Integration of hydrogenation and dehydrogenation based on dibenzyltoluene as liquid organic hydrogen energy carrier. Int J Hydrogen Energy. 2019;44:5345–54.
5. Asif F, Hamayun MH, Hussain M, Hussain A, Maafa IM. Park Y-K. Performance analysis of the Perhydro-Dibenzyl-Toluene dehydrogenation System—A simulation Study. Sustainability 2021.

Chapter 7
Comparison and Future Perspectives

Abstract The techno-economic assessment of H_2 transport through liquefied hydrogen, ammonia, toluene and dibenzyltoluene is presented in this chapter for the given bases of design. A sensitivity analysis is performed, considering different applications of the delivered H_2 (industrial and mobility sectors), different harbour-to-harbour distances and different cost of utilities and raw materials, identifying a present and a future scenario. The choice of the most cost-effective carrier is made based on the particular case under consideration. Key problems and limitations are critically identified to guide future large-scale implementation.

Keywords Liquefied hydrogen · Ammonia · LOHC · H_2 valley · HRS · H_2 value chain

7.1 Introduction

Discussed the layout of the overall H_2 transport value chain, a sensitivity study on the hypotheses on which the economic assessment is based on can be useful to understand their impact on results. Specifically, the sensitivity assessment considers:

- variable harbour-to-harbour H_2 transport in the range 2500–10000 km;
- different H_2 applications, namely H_2 valley and Hydrogen Refuelling Stations;
- variable utility costs, present and future, as presented in Chap. 2.

In this way, results are rationalized and the most cost-effective option is identified according to the discussed scenario.

7.2 H$_2$ Valley Case

Figure 7.1 shows the results of the techno-economic evaluations carried out, as a function of the distance D travelled by ship and for the two present and future scenarios.

The main cost driver in the LOHC value chain is the dehydrogenation. To lower the costs associated with transporting hydrogen via LOHC, it is essential to focus on the process intensification of this stage. For the ammonia value chain, the most impactful item in the overall economics is the synthesis of NH$_3$. To use ammonia as a hydrogen carrier cost-effectively on a smaller scale, enhancing the synthesis process is necessary. In the case of liquefied hydrogen, the highest expense comes from the liquefaction process, which needs further optimization to reduce electricity consumption.

For a transport distance of 2500 km, ammonia and dibenzyltoluene are the most cost-effective options for hydrogen transport and storage. However, at a distance of 5000 km, ammonia becomes more advantageous as a hydrogen carrier. This trend reversal occurs because, as the shipping distance increases, the initial loading cost of the organic carrier also rises, which is significant. At a distance of 10000 km, the cost of transporting hydrogen via LOHC becomes comparable to that of liquefied hydrogen due to the high cost of the initial organic carrier loading. Liquefied hydrogen remains the most expensive option for hydrogen transport due to its high liquefaction costs.

In future scenarios, the levelized cost of hydrogen ($LCOH$) varies depending on the carrier. Specifically, for liquefied hydrogen, the cost of liquefaction is expected to decrease significantly as the process becomes almost entirely powered by electricity, although the overall trend remains the same. For the LOHC value chain, the future LCOH is projected to be lower than it is currently, given the anticipated reduction in the cost of dibenzyltoluene.

7.3 HRS Case

To compare the different value chains of hydrogen transport in the HRS case, it has been assessed, for ammonia and liquefied hydrogen, where it is more cost-effective to carry out the reconversion, whether at the harbour (centralized reconversion) or at the end users (decentralized reconversion).

Figure 7.2 represents the result of the comparison between the centralized and decentralized conversion, for both liquefied hydrogen (Fig. 7.2a) and ammonia (Fig. 7.2b).

For liquefied hydrogen, decentralizing the reconversion to the end user is more convenient. In fact, hydrogen arrives at the user in its liquid state and can be pumped to the HRS operating pressure before vaporization, to save compression work downstream of regasification.

7.3 HRS Case

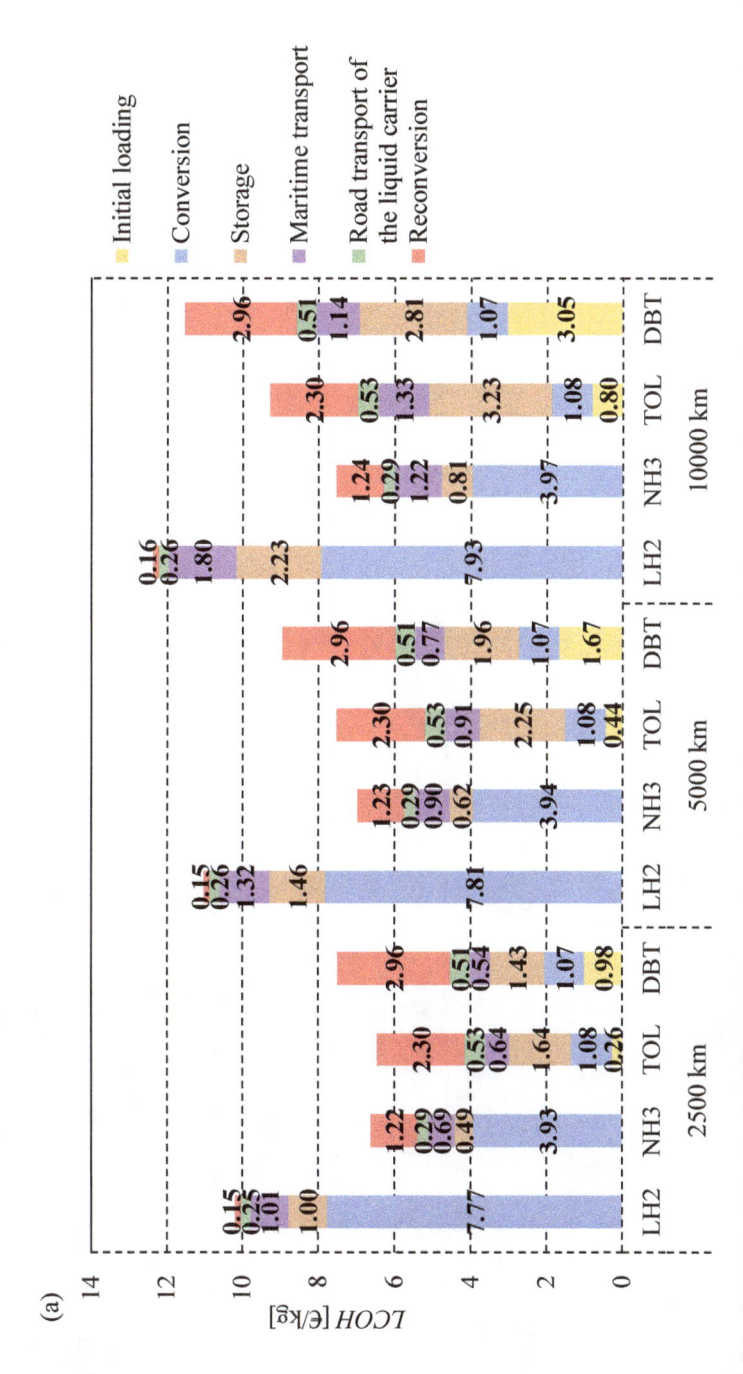

Fig. 7.1 Comparison between the different hydrogen transport value chains for its industrial application (H_2 valley), as a function of the distance D travelled by ship, for the scenarios: (**a**) present and (**b**) future

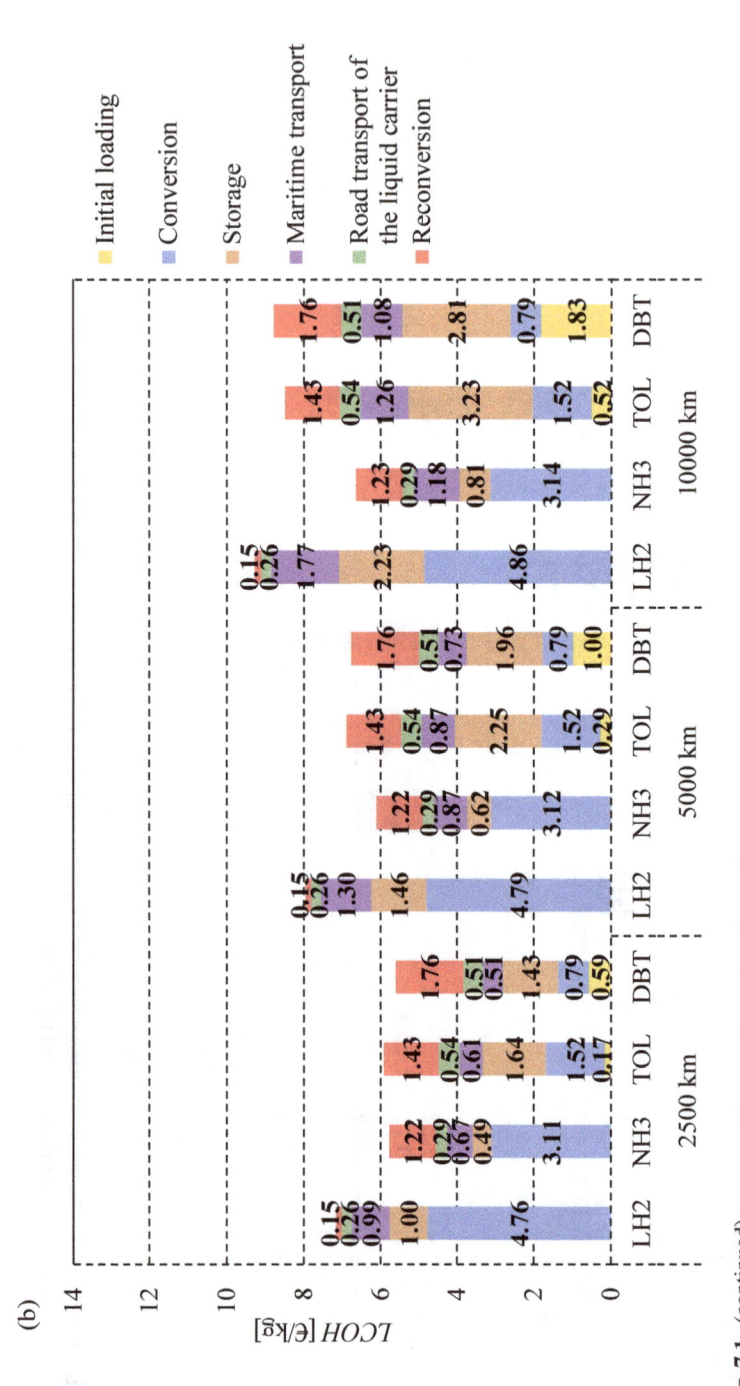

Fig. 7.1 (continued)

7.3 HRS Case

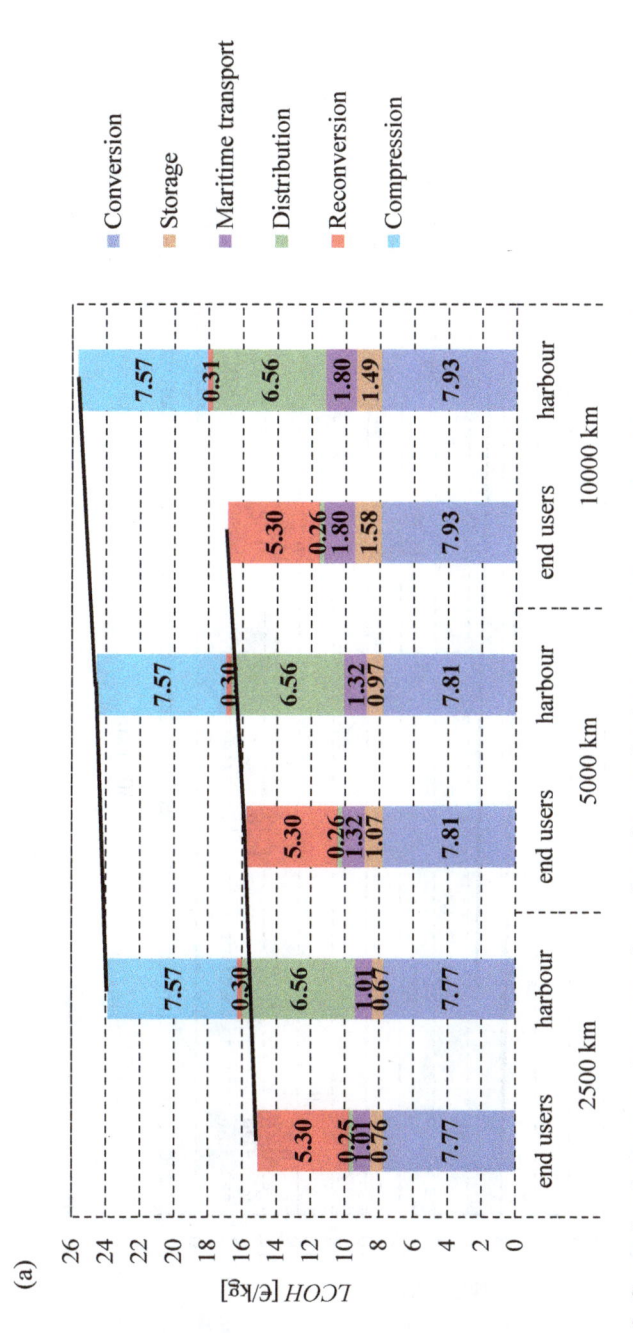

Fig. 7.2 Comparison between carrier reconversion at the end user and at the port in the HRS case (Fig. 2.2), as a function of the distance D travelled by ship, for the value chain: (**a**) liquefied hydrogen and (**b**) ammonia

74   7 Comparison and Future Perspectives

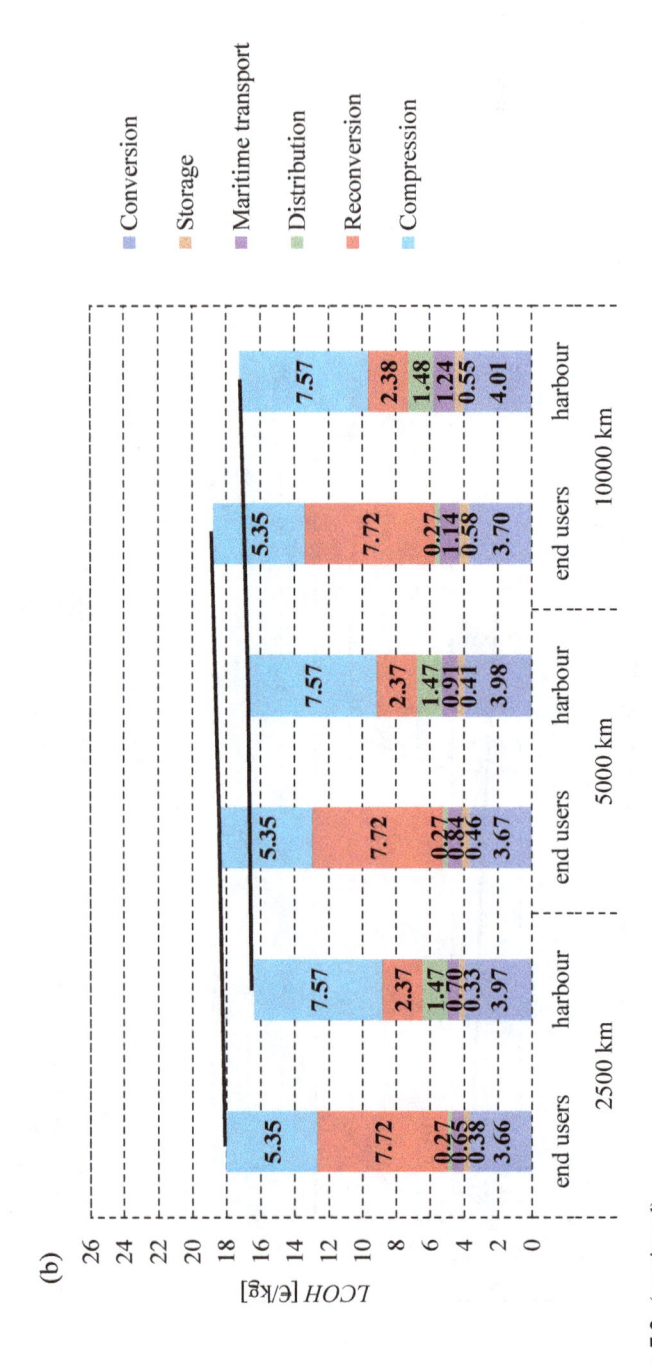

Fig. 7.2 (continued)

For ammonia, on the contrary, centralized cracking at the port of arrival, distribution of compressed hydrogen gas and subsequent recompression at the individual refuelling station is more economical. In fact, the cost of small-scale cracking, powered by electricity, is higher than the sum of centralized compression up to 320 bar and decentralized compression from 15 to 900 bar.

Having identified the best alternative, Fig. 7.3 shows the results of the techno-economic evaluations carried out, as a function of the distance D travelled by ship and for the two present and future scenarios. In all cases, liquefied hydrogen represents the most convenient alternative for the transport and storage of hydrogen, followed by ammonia and LOHC. The sensitivity analysis as a function of distance did not reveal any reversal of trend in this case. Liquefied hydrogen is the best carrier if hydrogen at 450/900 bar is required by small end consumers. This is due to the convenience that exists in transporting hydrogen in its liquid state and, therefore, in the possibility of pumping the liquid to save the compression work. Ammonia shows a lower cost than LOHCs. For both carriers, the reconversion takes place centrally at the port of destination, followed by compression up to 320 bar for the distribution of the gaseous hydrogen. In the future scenario, the cost difference between the different carriers is less evident since, as the price of electricity decreases, the costs for compression become less significant. Liquefied hydrogen is still the best alternative, while ammonia and LOHC show approximately the same cost.

7.4 Comparison with the Literature

The economic evaluations carried out for the cost drivers of the value chain of H_2 carriers are compared with what is available in the literature.

Table 7.1 reports the *LCOH* values obtained in this study and those reported in the literature for the processes involved in the liquefied hydrogen value chain. The estimates are significantly affected by the inflation (assumed base year for the economic assessment), H_2 capacity and electricity cost. For this reason, the values of these items are reported together with the corresponding estimates.

Figure 7.4 shows the extrapolation, obtained using the Hill's law, of the hydrogen liquefaction cost as a function of the plant's size for different values of the electricity cost. In particular, the solid blue curve corresponds to the present scenario (electricity cost of 500 €/MWh), while the solid orange curve represents the future scenario (electricity cost of 220 €/MWh). The dots indicate the liquefaction costs associated with the capacity being examined in this analysis. The cost of liquefying hydrogen is notably impacted by the price of electricity, resulting in a decrease of around 3 €/kg for liquefaction costs in the future scenario with respect to the present one. The crosses represent the costs of hydrogen liquefaction as reported in existing literature. Due to the impact of electricity prices, the curves representing the calculated liquefaction costs as function of plant capacity for the different electricity cost values are plotted in colours corresponding to the sources of those electricity cost values. In particular, the electricity price of 100 €/MWh, which corresponds to the violet dashed line, is

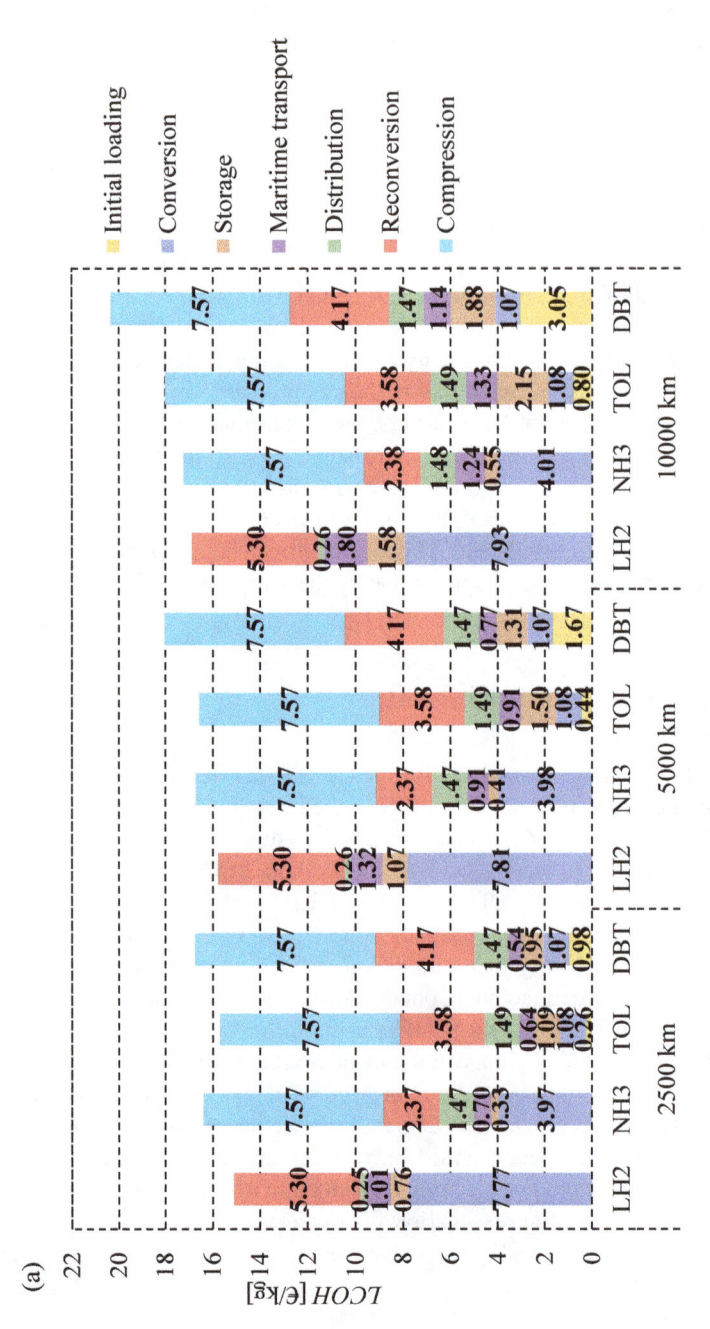

Fig. 7.3 Comparison between the different hydrogen transport value chains for its application in the mobility sector (HRS), as a function of the distance D travelled by ship, for the scenarios: (**a**) present and (**b**) future

7.4 Comparison with the Literature

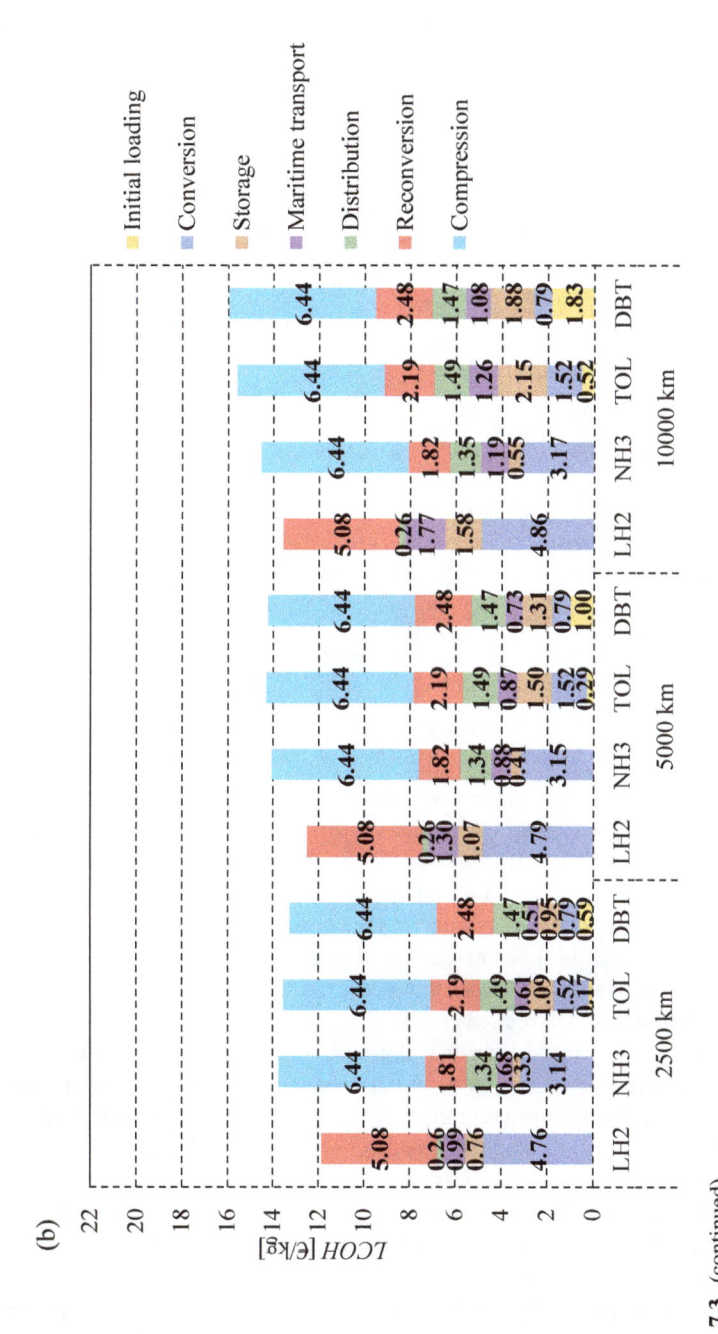

Fig. 7.3 (continued)

Table 7.1 Costs for the LH$_2$ liquefaction, centralized regasification and decentralized regasification: comparison between the carried-out assessments and the data available in the literature

References	Base year	H$_2$ capacity [tpd]	Electricity cost [€/MWh]	LCOH [€/kg] conversion	LCOH [€/kg] centralized reconversion	LCOH [€/kg] decentralized reconversion
This study present scenario	2022	43	500	7.77	0.15	5.30
This study future scenario	2022	43	220	4.76	0.15	5.08
Chodorowska and Farahadi [1]	2020	500	35*	0.834[a]	0.04[a]	–
Hank et al. [2]	2020	120	25	0.43	–	–
IEA [3]	2019	700	98[a]	1.0[a]	0.05[a]	0.05[a]
Ishimoto et al. [4]	2015	500	38	2.75–3.13	–	–
Niermann et al. [5]	2020	500	80	2.0	0.6	–
Okunlola et al. [6]	2020	607	58[a]	2.53*	0.60–1.35[a]	–
Raab et al. [7]	2018	676.5	100	1.76	0.30	–
Berger [8]	2025	200	–	0.9	0.04	–

[a]Currency is US $, instead of €

taken from Raab et al. [7] (liquefaction cost denoted by a violet plus symbol), while the electricity price of 25 €/MWh, which corresponds to the green dashed line, is obtained from Hank et al. [2] (liquefaction cost denoted by a green plus symbol). When examining Fig. 7.4, a good agreement can be seen between the estimates derived in this analysis and those found in the current body of literature.

As regards NH$_3$ synthesis and cracking, the calculated costs are compared with the available bibliography on the techno-economic analysis of the ammonia value chain and are reported in Table 7.2.

From Table 7.2 it can be seen that the reported costs have the same order of magnitude, even if a direct comparison between the different sources analysed is difficult, as the capacity and methodology used are significantly different. In all cases, however, centralized cracking has a cost per kg of hydrogen that is an order of magnitude lower than decentralized cracking.

As regards the decentralized reconversion, to evaluate the quality of the results obtained and better compare them with what is reported in the literature, the cracking cost is calculated for different capacities considering the *CAPEX* appropriately scaled through the Hill's law and the *OPEX* calculated starting from different prices for electricity: 500 €/MWh (blue curve in Fig. 7.5b), 220 €/MWh (orange curve in

7.4 Comparison with the Literature

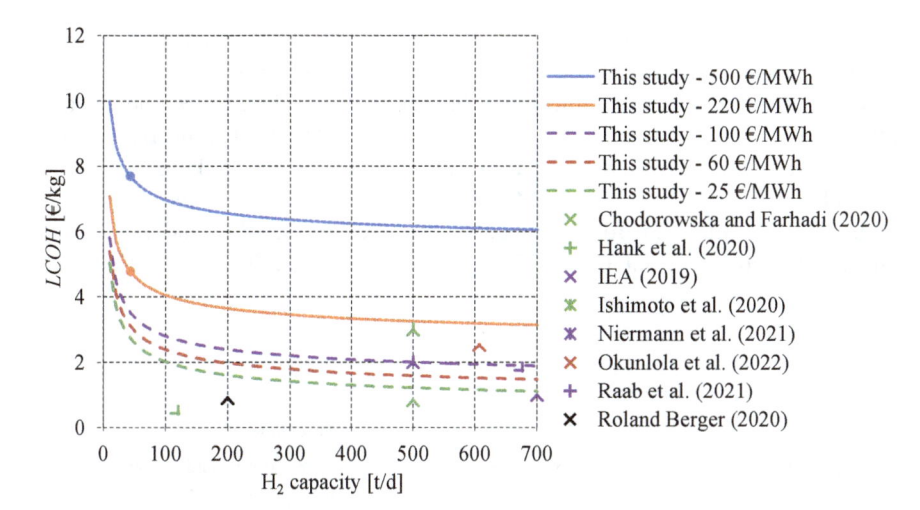

Fig. 7.4 LCOH [€/kg] for the hydrogen liquefaction process: comparison between the results obtained in this study and the literature [1–8]

Table 7.2 Costs for the NH_3 synthesis, centralized cracking and decentralized cracking: comparison between the carried-out assessments and the data available in the literature

References	Base year	H_2 capacity [tpd]	Electricity cost [€/ MWh]	LCOH [€/ kg] conversion	LCOH [€/ kg] centralized reconversion	LCOH [€/kg] decentralized reconversion
This study present scenario	2022	43	500	3.97	1.22	7.72
This study future scenario	2022	43	220	3.14	1.22	4.87
Chodorowska and Farahadi [1]	2020	500	35[a]	0.473[a]	0.427[a]	–
Hank et al. [2]	2020	120	25	0.3	–	–
IEA [3]	2019	700	98[a]	0.3[a]	0.8[a]	1.0[a]
Ishimoto et al. [4]	2015	500	38	4.29	0.93	–
Okunlola et al. [6]	2020	607	58[a]	1.36[a]	1.07[a]	–
Perna et al. [9]	2020	0.2	119–137	–	–	6.92–7.35
Berger [8]	2025	200	–	0.5	1.0	–

[a]Currency is US $, instead of €

Fig. 7.5b) and 100 €/MWh (as assumed by Perna et al. [9], violet dashed curve in Fig. 7.5b). For the centralized ammonia cracking (Fig. 7.5a), the electric power involved is not such as to significantly influence the cost. The *LCOH* calculated in this way is close to that reported in the literature for centralized reconversion, while it differs from the estimate of Perna et al. [9] for decentralized cracking.

Tables 7.3 and 7.4 report the *LCOH* values obtained in this study and those reported in the literature for the processes involved in the TOL and DBT value chains, respectively. From Tables 7.3 and 7.4 it can be seen that the reported costs

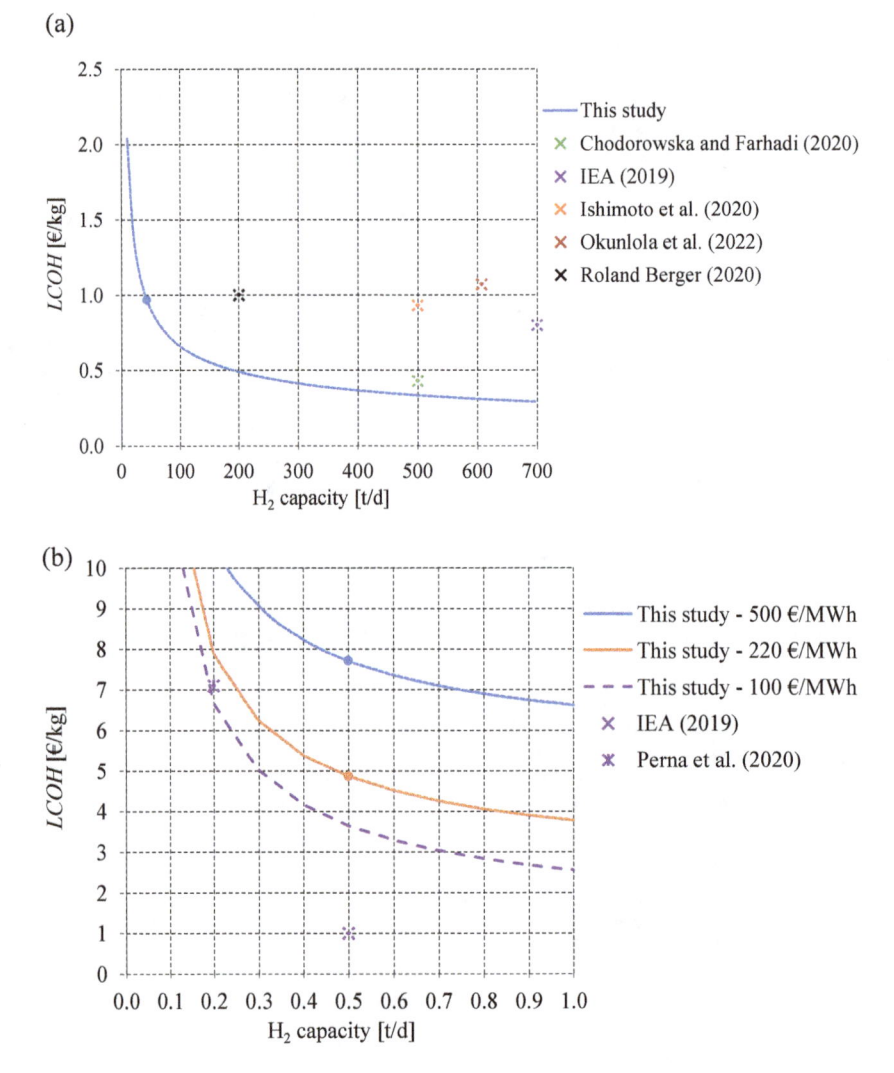

Fig. 7.5 *LCOH* [€/kg] for **a** centralized and **b** decentralized ammonia cracking process: comparison between the results obtained in this study and the literature [1, 3, 4, 6, 8, 9]

7.4 Comparison with the Literature

have the same order of magnitude, even if a direct comparison between the different sources analysed is difficult, as the capacity and methodology used are significantly different. In all cases, however, dehydrogenation shows higher costs than hydrogenation, suggesting that the latter is the real cost driver of the value chain.

Since dehydrogenation is the most significant cost item in the entire chain, a more detailed comparison is carried out, shown in Fig. 7.6. In particular, the *LCOH* is calculated for different capacities considering the *CAPEX* appropriately scaled

Table 7.3 Costs for the TOL hydrogenation and centralized dehydrogenation: comparison between the carried-out assessments and the data available in the literature

References	Base year	H_2 capacity [tpd]	Electricity cost [€/MWh]	LCOH [€/kg] conversion	LCOH [€/kg] centralized reconversion
This study present scenario	2022	43	500	1.08	2.30
This study future scenario	2022	43	220	1.52	1.43
Chodorowska and Farahadi [1]	2020	500	35[a]	0.339[a]	0.473[a]
IEA [3]	2019	700	98[a]	0.4*	1[a]
Niermann et al. [5]	2020	500	80	0.1	1.9
Raab et al. [7]	2018	676.5	100	0.81	1.22

[a]Currency is US $, instead of €

Table 7.4 Costs for the DBT hydrogenation and centralized dehydrogenation: comparison between the carried-out assessments and the data available in the literature

References	Base year	H_2 capacity [tpd]	Electricity cost [€/MWh]	LCOH [€/kg] conversion	LCOH [€/kg] centralized reconversion
This study present scenario	2022	43	500	0.98	2.96
This study future scenario	2022	43	220	0.59	1.76
Hank et al. [2]	2020	120	25	0.9	–
Niermann et al. [5]	2020	500	80	0.2	0.5
Raab et al. [7]	2018	676.5	100	0.48	0.97

through Hill's law and the *OPEX* calculated starting from different electricity prices: 500 €/MWh and 220 €/MWh (blue and orange solid lines in Fig. 7.6, respectively), 100 €/MWh (as assumed by Raab et al. [7], purple dashed line in Fig. 7.6), 25 €/MWh (close to the value assumed by Chodorowska and Farhadi [1], green dashed line in Fig. 7.6a). For toluene, the *LCOH* calculated in this way is close to that reported by Chodorowska and Farhadi [1], while for dibenzyltoluene the *LCOH* is similar to that reported by Raab et al. [7] and Niermann et al. [5] (see the curves in Fig. 7.6 compared with the crosses of the same colour).

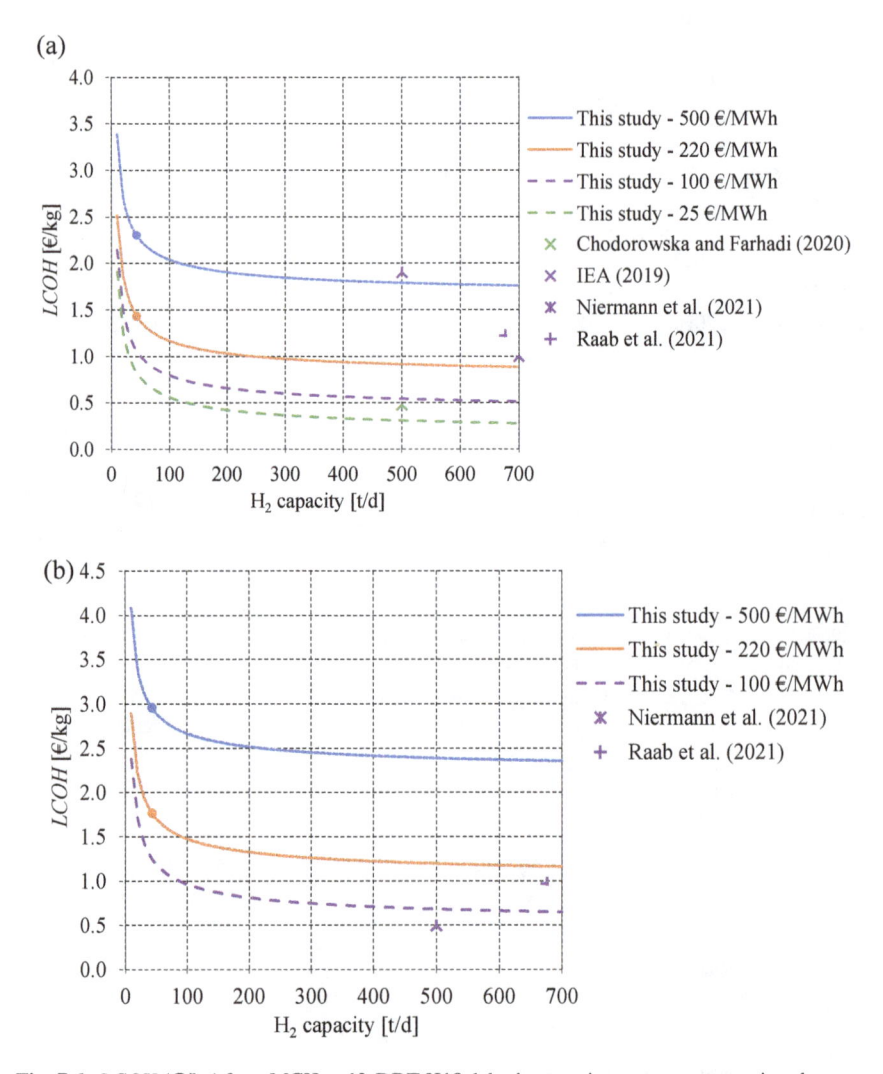

Fig. 7.6 *LCOH* (€/kg) for **a** MCH and **b** DBT-H18 dehydrogenation process: comparison between the results obtained in this study and the literature [1, 3, 5, 7]

7.5 Conclusions and Future Perspectives

The performed techno-economic analysis allowed to delve into each stage of the value chains of the different carriers for hydrogen transport and to discuss the possible application in the industrial sector (H$_2$ valley case) and in the mobility sector (Hydrogen Refuelling Station case). In general, the application to the industrial sector appears to be the one with the lowest costs compared to the mobility application, the latter being affected by the high impact of compression costs and the safety issues related to the management of gaseous hydrogen[1].

This study revealed both the advantages and the critical issues associated with each of the considered carriers. The comparison between the different carriers is shown in Table 7.5.

Worldwide, many projects are focusing on the development of a hydrogen economy. In Europe, the IPCEI Hy2Use project, involving 29 companies from 13 states, aims at first industrial deployment and construction of relevant infrastructure in the hydrogen value chain. Several projects are expected to be implemented in the near future, with various large-scale electrolysers expected to be operational by 2024–2026 and many of the innovative technologies deployed by 2026–2027 [10].

Table 7.5 Pros/cons of the different hydrogen carriers analysed

Carrier	Pros	Cons
LH$_2$	✓ Only electric energy needed for the liquefaction process	✗ Liquefaction is an expensive process, due to the high electricity consumption;
	✓ Cost-effective H$_2$ carrier for mobility application (specifically in the future scenario)	✗ Ad-hoc infrastructure needed
NH$_3$	✓ Cost-effective H$_2$ carrier for industrial application at high harbour-to-harbour distances	✗ Toxicity issues;
	✓ Highly flexible (it can be used both as H$_2$ carrier and carbon-free energy vector)	✗ Low TRL of the cracking process;
	✓ Already a global commodity transported internationally by ship and pipeline	✗ Need for cost reduction of the small-scale synthesis process
LOHC	✓ Cost-effective H$_2$ carrier for industrial application at low harbour-to-harbour distances	✗ Low TRL → the technology has to be proved at the industrial scale;
	✓ Effective for long-term H$_2$ storage without losses	✗ High cost of LOHC, mainly DBT, and uncertainty of DBT market;
	✓ Can be easily handled with the existing infrastructure for oil products	✗ Need for cost reduction of the dehydrogenation process

---

[1] https://doi.org/10.1016/j.ijhydene.2024.07.241

The completion of the overall project is planned for 2036. Saudi Arabia's NEOM Green Hydrogen Company is constructing a plant that will integrate up to 4 GW of solar and wind energy to produce up to 600 t/d of carbon-free hydrogen by the end of 2026, to be stored and shipped in the form of ammonia as a cost-effective solution for the transportation and industrial sectors globally [11]. Australia made history by exporting liquefied hydrogen to Japan in 2022 through the Hydrogen Energy Supply Chain (HESC) pilot project [12]. LATTICE Technology and LH2 Energy are advancing a liquefied hydrogen export chain from Darwin (Australia) to Korea. They are currently conducting the basic design project for 75000 m^3 liquefied hydrogen export and import terminals [12]. New Zealand's Meridian has selected Australia's Woodside Energy as its partner to develop a large-scale hydrogen and ammonia facility at Southland (New Zealand). Japan's Mitsui is also in talks to join the proposed project, dubbed Southern Green Hydrogen (SGH), that is targeting to produce 500000 t/y of ammonia using electrolysis from renewable power [13]. In the U.S., there are 18 blue hydrogen projects, representing a total of 22000 t/d of hydrogen capacity. Several green hydrogen projects are also in development, although most are small pilot plants, with a cumulative total of 12000 t/d of hydrogen capacity [14]. While new projects are spread across the country, there is a significant concentration planned along the Gulf Coast, accounting for about 65% of the planned capacity. Some of these projects will supply demand at existing industrial complexes. However, many have indicated that their products will be sold in export markets or used as transportation fuels. Among the ten largest hydrogen projects planned for the Gulf Coast, four are intending to produce methanol or methanol-based fuels for transportation markets. Another four are planning to produce ammonia, with most targeting export markets, especially in Asia, as their primary source of demand.

In conclusion, the global shift towards a hydrogen economy is evident from the multitude of projects underway worldwide. The planned large-scale hydrogen production facilities, coupled with the commitment to renewable energy sources [15, 16], indicate a strong momentum towards sustainable hydrogen production. Moreover, the focus on export markets underscores the potential of hydrogen carriers, such as liquefied hydrogen, ammonia, and liquid organic hydrogen carriers, as cost-effective solutions for hydrogen transport. As these projects continue to develop and innovate, they pave the way for a cleaner, more sustainable future powered by hydrogen.

References

1. Chodorowska N, Farhadi, M. H2 value chain comparing different transport vectors. In: GPA Europe Virtual Conference, 25 May 2021.
2. Hank C, Sternberg A, Köppel N, Holst M, Smolinka T, Schaadt A, et al. Energy efficiency and economic assessment of imported energy carriers based on renewable electricity. Sustainable Energy Fuels. 2020;4:2256–73.
3. IEA. The Future of Hydrogen;2019.

References

4. Ishimoto Y, Voldsund M, Nekså P, Roussanaly S, Berstad D, Gardarsdottir SO. Large-scale production and transport of hydrogen from Norway to Europe and Japan: Value chain analysis and comparison of liquid hydrogen and ammonia as energy carriers. Int J Hydrogen Energy. 2020;45:32865–83.
5. Niermann M, Timmerberg S, Drünert S, Kaltschmitt M. Liquid Organic Hydrogen Carriers and alternatives for international transport of renewable hydrogen. Renew Sustain Energy Rev. 2021;135:110171.
6. Okunlola A, Giwa T, Di Lullo G, Davis M, Gemechu E, Kumar A. Techno-economic assessment of low-carbon hydrogen export from Western Canada to Eastern Canada, the USA, the Asia-Pacific, and Europe. Int J Hydrogen Energy. 2022;47:6453–77.
7. Raab M, Maier S, Dietrich R-U. Comparative techno-economic assessment of a large-scale hydrogen transport via liquid transport media. Int J Hydrogen Energy. 2021;46:11956–68.
8. Berger R. Hydrogen transportation;2021.
9. Perna A, Minutillo M, Di Micco S, Cigolotti V, Pianese A. Ammonia as hydrogen carrier for realizing distributed on-site refueling stations implementing PEMFC technology. In: E3S Web of Conferences, vol. 197. EDP Sciences;2020. p. 05001.
10. https://ec.europa.eu/commission/presscorner/detail/en/ip_22_5676.
11. https://www.neom.com/en-us/newsroom/neom-green-hydrogen-investment.
12. https://www.hydrogenenergysupplychain.com/.
13. https://www.energyvoice.com/renewables-energy-transition/464476/woodside-and-mitsui-join-new-zealand-green-hydrogen-export-project/.
14. https://insight.factset.com/u.s.-hydrogen-projects-tilt-toward-export-markets-adding-dem and-for-natural-gas.
15. Moioli S, Pellegrini LA. Operating the CO_2 absorption plant in a post-combustion unit in flexible mode for cost reduction. Chem Eng Res Des. 2019;147:604–14.
16. Moioli S, Pellegrini LA. Fixed and capture level reduction operating modes for carbon dioxide removal in a Natural Gas Combined Cycle power plant. J Clean Prod. 2020;254:120016.